BUILDING CONSTRUCTION

Interior Systems

BUILDING CONSTRUCTION

Interior Systems

James Ambrose
University of Southern California

VNR VAN NOSTRAND REINHOLD
_____ New York

Copyright © 1991 by Van Nostrand Reinhold

Library of Congress Catalog Card Number 90-45718
ISBN 0-442-00292-0

Manufactured in the United States of America

Published by Van Nostrand Reinhold
115 Fifth Avenue
New York, New York 10003

Chapman and Hall
2-6 Boundary Row
London, SE 1 8HN

Thomas Nelson Australia
102 Dodds Street
South Melbourne 3205
Victoria, Australia

Nelson Canada
1120 Birchmount Road
Scarborough, Ontario M1K 5G4, Canada

16   15   14   13   12   11   10   9   8   7   6   5   4   3   2   1

Library of Congress Cataloging-in-Publication Data
Ambrose, James E.
    Building construction : interior systems / James Ambrose.
        p.   cm.
    Includes index.
    ISBN 0-442-00292-0
    1. Building—Details.   I. Title.
TH2025.A43   1991
    690.1—dc20   90-45718
                                                         CIP

Contents

Preface

This book treats a subject that is of interest to a wide range of people with various relationships to the designing and constructing of buildings. There is at any given time a great mass of information on these topics which may be accessed for application to the many tasks of designers, builders, suppliers, and others. The enormity of this information resource is at once reassuring to those who regularly encounter needs for it, and overwhelming to just about everyone who needs to figure out how to use it.

This book relates the topic of building construction to the basic problems of building design. The basic assumption here is that the need to design precedes the need to build, and that real concern for how to build comes from a desire to build something. This is the normal process of development for designers and others who start from the point of desiring a building and then proceed to determine what it should be. Intense concern for specific consideration of building materials, systems, and details of construction thus emerges at a later stage of design, typically after the general form, size, and essential nature of the building are already proposed.

A major problem with using the great mass of available information about building construction is that it is largely not oriented to the purposes of education or design; thus it is not user-friendly to the student or apprentice in architecture, or to any others who do not already have a broad grasp of what buildings and building construction are all about. Rich as they may be as information sources, *Sweets Catalog Files*, the *CSI Spec-Data System*, and *Architectural Graphic Standards* are not friendly to inexperienced users. You have to already know the trade lingo, what you are looking for, and pretty much why you are looking for it to make effective use of these resources.

The latest, most extensive, most detailed information about building construction is largely produced by persons engaged in the manufacturing, supplying, specifying, purchasing, regulating, financing, or insuring of building materials and products, and in their applications to making buildings. The information presented is often slanted toward the specific concerns of those who produce it. Thus it is to be expected that materials forthcoming from manufacturers and suppliers are shaped to the purpose of promoting sales of their products, that insurers and financiers have strong economic concerns, and that specification writers and enforcers of building codes are somewhat paranoid about specific language and terminology that is legally binding.

This book attempts to be user-friendly to the person who is basically more interested in buildings in general and somewhat less in the specific concerns for their construction. The basic purposes here are to develop a general view of buildings and the problems of their design as they relate to the eventual need to construct them. Highly specific, detailed information, such as how to attach gypsum drywall to wood studs, can be pursued through appropriate sources, once the real need for the information is established. The need for having drywall, for attaching it to the studs, for having wood studs—and indeed, for having the wall in the first place—need to precede the search for the specific information.

Buildings are complex and the topic of their construction is correspondingly extensive. In order to cut down the size of this presentation, the topic here is limited to concerns for the development of the building interior, consisting primarily of floors, ceilings, interior partitions, and stairs. That is what most of us know from the direct experience of the buildings we use, since what we mostly use is the building's enclosed spaces. It is also what building designers develop most specifically in relation to the needs of the building users. Nevertheless, it leaves a lot of the building not treated. This book is, in fact, the second volume of a series. The first volume treated the subject of the building's basic shell that forms the enclosure, consisting primarily of roofs, exterior walls, windows and doors. I expect to pursue the topic in subsequent volumes to consider building sites and building service systems.

My views of the need for this book and my ideas for its contents have emerged from many years of experience as a building designer and a teacher. I am grateful to all my former students, co-workers, and others whose feedback of frustrations have shaped those views and ideas. I am also grateful to the many people at Van Nostrand Reinhold who have helped to bring my rough materials into being as an actual book; particularly my editors Everett Smethurst and Wendy Lochner as well as Cindy Zigmund and Alberta Gordon.

I am also grateful to various organizations who have permitted the use of materials from their publications, as noted throughout the book.

Finally, I am grateful to the members of my family who steadfastly support and tolerate me in my home office working situation. For this book, I am particularly grateful for the assistance of my wife, Peggy, and my son, Jeffrey.

James Ambrose

BUILDING CONSTRUCTION

Interior Systems

1

Introduction

Design of building interiors is a major area of concern in the general design of buildings. This chapter presents discussions of the issues and specific problems of building interiors and how the topic is developed in this book.

1.1 DESIGN OF INTERIORS: GENERAL CONSIDERATIONS

In its broadest sense, interior design is a problem with many aspects, ordinarily involving the concerns of many designers (see Figure 1.1). The general form and appearance of interiors is of major concern to architects, but also to interior designers and to many other persons who deal with the shaping, finishing, and furnishing of interiors of buildings.

While appearance and general spatial functioning of interiors is a critical concern, there are many other design considerations for interiors, including the following major ones:

Lighting, both natural and artificial

Sound, for privacy and general noise control

Thermal comfort

Air quality

Fire safety

Security for people and contents

The shaping and general constructing of building interiors must respond to all of these concerns—and, in many situations, to various additional ones.

The primary topic of this book is not interiors in general, but rather their basic construction. However, the selection of materials, components, systems, and details for interior construction impinges on the work of the designers of all of the aspects of the building interior. While it is possible to consider only a few concerns, or even a single one, all of the potential effects of choices for the construction must eventually be dealt with.

While the building interior is a definable place and a specific design concern, it must also relate to the general building enclosure, the building structure, and the many building service systems. In the end, the whole building must be designed, even though specific interior design situations may be dealt with as individual problems.

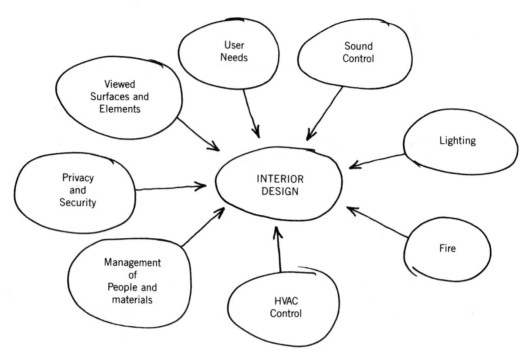

Figure 1.1   Scope of concerns in interior design.

This book deals specifically with problems of constructing the building interior, which indeed is a part of the building design. Relations to many other parts of the building design are discussed, but the primary focus is the design for interior accommodation.

The point of view here is that of the designer and what is of prime concern in the design process. The writing of specifications, management of construction, and production of materials and components for construction are necessary activities that will be considered, but design is the focus.

1.2 BUILDING CONSTRUCTION AS A DESIGN PROBLEM

The making of buildings is a big business, affecting many people. However, the development of the materials in this book is based, in part, on the attitude that the determination of building construction is an architectural design problem. The solving of problems, large and small, with regard to the construction work affects the design—or at least it ought to. If certain details or dimensions are not feasible to achieve with a particular material or system, the building designers should accept the facts, or prove by some means that what is proposed for the design can truly be accomplished. Design work must be informed, not necessarily in early sketching stages, but as soon as any serious commitments are made and the detailed development of the work proceeds.

It is possible, of course, to make the building construction itself a major design determinant. However, respect for correctness and expression of the construction is one thing; reverence for it is another. Many architects are more concerned with form, space, appearance, illusion, metaphor, symbolic relations, human response, or other matters, caring somewhat less about the expression of the construction or the revealing of the building's functioning elements.

For most designers the actual work of design usually is a very personal matter. Launching points, decision sequences, and the relative weights of different values vary considerably from designer to designer as well as from project to project with a single designer. Approaching the subject of building construction as an architectural design issue simply means that concerns for the construction emerge as necessities in the process of design. The designer does not set out to produce some construction, but rather encounters construction considerations in the normal process of designing a building. First comes the concept of the building and all of its planning concerns— then the problems of making it.

There is no intention to advocate any particular design philosophy in this book. Proper construction is considered essentially as a simple, pragmatic matter. It is considered to have a reasonable value and to be something that deserves serious attention—but it is not essential that it command dominance in all architectural design decisions.

1.3 HOW TO USE THIS BOOK

The basic purpose of this book is to help the inexperienced designer, whether in a school design class or in an apprentice position in a design office. However, anyone who is interested in the topic of building construction and its effects on building design should be able to derive some use from this book.

There is no single method for the most effective use of the materials in this book. Readers are sure to vary considerably with regard to interests, needs, and backgrounds. While the materials have been arranged with some logical sequence in mind, it is anticipated that the typical reader will take up the materials in a random fashion. This is to be expected all the more if the reader is seeking help in the progress of some design work.

We do not mean to discourage any reader from taking up the book materials in the sequence presented. This is likely to make most sense, if the reader is essentially uninformed in the general topic. Development of a fluent "construction" vocabulary is also to be generally encouraged, as the discipline of proper terminology is quite essential for accurate communication in the building design and construction business world.

Finally, for many readers, this book may serve a principal function in pointing out sources for additional information. General major sources are discussed in the next section and various special sources are cited throughout the book in the discussions of specific topics.

1.4 SOURCES OF DESIGN INFORMATION

Information about building construction is forthcoming from many sources and relates to many potential uses (see Figure 1.2). For the specific purpose of supporting design activity, some particular sources and applications are of special value. This is a highly variable situation, responding to many considerations for a particular design activity. Significant variables include the following:

Stage of the design process—from very preliminary to final construction directives

Nature of the design project—large or small; luxury or low cost; specific constraints (codes, special occupant problems, etc.)

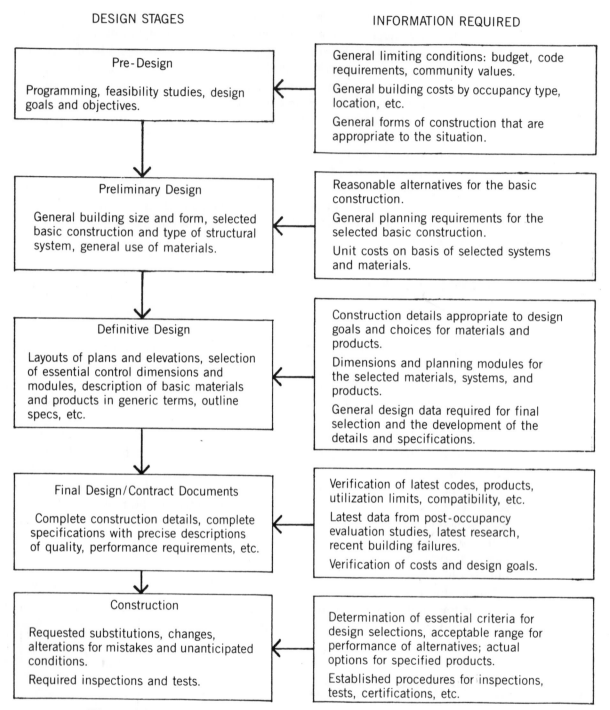

DESIGN STAGES

INFORMATION REQUIRED

Pre-Design

Programming, feasibility studies, design goals and objectives.

General limiting conditions: budget, code requirements, community values.

General building costs by occupancy type, location, etc.

General forms of construction that are appropriate to the situation.

Preliminary Design

General building size and form, selected basic construction and type of structural system, general use of materials.

Reasonable alternatives for the basic construction.

General planning requirements for the selected basic construction.

Unit costs on basis of selected systems and materials.

Definitive Design

Layouts of plans and elevations, selection of essential control dimensions and modules, description of basic materials and products in generic terms, outline specs, etc.

Construction details appropriate to design goals and choices for materials and products.

Dimensions and planning modules for the selected materials, systems, and products.

General design data required for final selection and the development of the details and specifications.

Final Design/Contract Documents

Complete construction details, complete specifications with precise descriptions of quality, performance requirements, etc.

Verification of latest codes, products, utilization limits, compatibility, etc.

Latest data from post-occupancy evaluation studies, latest research, recent building failures.

Verification of costs and design goals.

Construction

Requested substitutions, changes, alterations for mistakes and unanticipated conditions.

Required inspections and tests.

Determination of essential criteria for design selections, acceptable range for performance of alternatives; actual options for specified products.

Established procedures for inspections, tests, certifications, etc.

Figure 1.2   Use of construction information in the building design process.

Working context of the designer—time constraints; design budget; size of the design team; access to information sources

Individual types of information and types of information sources may be of particular value, depending on the design work as qualified by the preceding, as well as by many additional considerations. Some major categories of information sources that are generally supportive of design work are the following:

Books—including ones of broad scope, such as general references on building construction, and ones of a highly specialized nature with important data for a very specific problem

Industry-supplied materials—mostly supplied for promotional purposes, but nevertheless usually indispensable for specific product data

Records of designs—statistical, graphic, or photographic records of proposed or actual construction

For any specific design work, the truly usable or necessary references may relate as much to the particular needs of the individual designer as to the specific nature of the design activity. The education, training, developed skills, and previous experience of the designer may well make certain general references indispensable or generally require help only for very special situations.

This book is not intended as an inexhaustible source of specific information about all of building construction. Although the view is broad, the scope of the presentations here is narrowly bound by specific concentration on the needs of the designer, the general nature of design work, and particular interest in building interiors. It is generally assumed that the reader—whether in a school or in a design office—has access to some general information sources. These include the following basic items:

One or more general texts on basic building construction materials, products, and processes

One or more general references on standard details of building construction

Some industry-supplied information; as generally represented by *Sweets Catalog Files* or a selected, personally assembled set of individual entries as obtainable from individual manufacturers or agencies

It is assumed that perceptive readers are aware of some of the pitfalls in using information sources. We refer particularly to the major concerns for timeliness, neutrality of the suppliers, and feasible applicability to the design work at hand. In the latter regard are such concerns as local codes, availability of products and services, climate differences, and so on.

Building construction is largely achieved with commercially developed products. The suppliers of these products do a lot of hard selling to compete in the marketplace, and information obtained from them is quite understandably not of a neutral, objective nature. You can't expect to be able to judge the relative appropriateness of a product to a specific application on the basis of the information supplied by the product's supplier.

Many references have been used in the development of the work presented in this book. Many of these sources are cited for very specific information, but the general development of the work derives often from a collective consideration of many sources. For general reference, we acknowledge—and recommend to the reader—the following sources:

1. *Architectural Graphic Standards,* 8th ed., Ramsey and Sleeper, New York: Wiley, 1988.
2. *Fundamentals of Building Construction: Materials and Methods,* 2nd ed., Edward Allen, New York: Wiley, 1990.

3. *Construction Materials and Processes,* 3rd ed., Don Watson, New York: McGraw-Hill, 1986.
4. *Construction Principles, Materials, and Methods,* 5th ed., Olin, Schmidt, and Lewis, New York: Van Nostrand Reinhold, 1983.
5. *Sweet's Catalog Files: Products for General Building and Renovation,* New York: McGraw-Hill, annually published.
6. *Mechanical and Electrical Equipment for Buildings,* 7th ed., Stein, Reynolds, and McGuinness, New York: Wiley, 1986.

For specific problems and applications, many additional references are cited throughout the book.

2

Architectural Components

This chapter presents materials relating to the most ordinary and generally indispensable parts that occur in the majority of buildings and with some of the design concerns that arise in producing them. While designers are necessarily concerned with all of a building's parts, the concentration here is on those parts most directly involved in achieving the building's interior.

Design of any selected building part must be done in the context of the problems of the whole building. Dealing with the individual elements of building construction requires some attention to their eventual coordination in the whole constructed building. Putting the parts together is the theme of the work in Chapter 7, where the whole construction of several buildings is presented.

The objective in this chapter is to deal generally with the various concerns for the basic building component parts. For real situations, consideration must be given to the means for producing the parts in terms of available materials and products. Chapter 3 presents discussion of basic construction materials. Chapter 4 examines the various products used to achieve basic building components. Chapter 5 discusses some special, although highly crucial, concerns for interiors. Chapter 6 considers the general problems of systems.

2.1 INDIVIDUAL COMPONENTS

For the purpose of discussion, we will consider a two-level set of basic architectural components. On the first level are those major elements that principally determine the structure and basic character of the construction of the building. As shown in Figure 2.1, these major elements are the roof, floors, exterior walls, foundation elements, and elements for vertical circulation (stairs, elevators, etc.).

On the second level are elements which, although necessary, have generally less influence on determination of the basic form of the structure or the choice of the principal materials and methods of construction. The second level includes windows, doors, ceilings, and various items involved in the building service systems (for power, lighting, HVAC, etc.). The second level may also contain site features outside the building.

In any given situation one or more of the second-level elements may take a dominant role in the set of values for design. For example, in an office building, the need for development of an integrated ceiling system may strongly constrain the

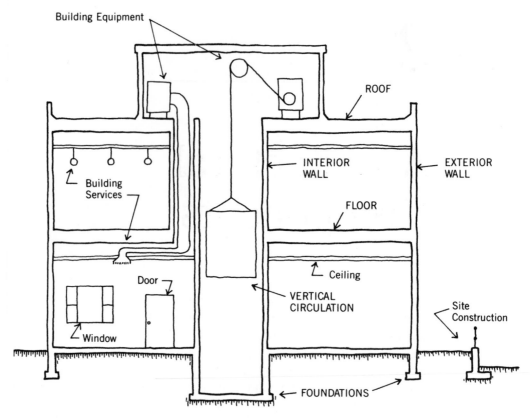

Figure 2.1   Fundamental construction components for buildings.

design of the roof or floor structure, interior partitioning, and services for lighting and ceiling-mounted HVAC elements, fire sprinklers, intercom systems, and so on.

Staying close to the central theme of this book, we will limit the discussion here to those elements with primary relationships to the functions of the building interior. This includes first-level components of floors, interior walls, and vertical circulation and second-level components of ceilings and interior doors.

In the end, of course, the whole building must be dealt with and the interior cannot be developed without consideration of the building equipment and services and the relations of the interior to the building enclosure. Many of these relations will indeed be considered in the discussions that follow, although the full treatment of these other parts of the building will not be developed here.

2.2 FLOORS

Floors are working platforms and must respond to the needs of the building users. Ordinarily, this means the desire for a dead flat surface and an exposed finish that responds to practical concerns for the type of traffic, anticipated maintenance, budget, and various functional demands deriving from concerns for fire, sound control, and so on.

Floor finishes endure considerable wear and soiling and are often replaced or reconditioned with some frequency. For this reason, finishes are often dealt with as applied surfacing that can be removed and replaced or restored without damage to the supporting structure. The supporting structure must facilitate the installation— as well as the replacement—of the applied finish, and should usually be able to facilitate some changes in future types of finishes.

There are two major categories of supporting floor structures: those placed directly on the ground, and those that span over open space beneath them (called framed floors). Floors on the ground are most often achieved with concrete pavements. Framed floors can be achieved with a wide variety of structures, depending on the length of spans, fire requirements, and the materials of the general construction of the building.

When floors are framed, they become separating elements between spaces. This brings three major relationships into consideration. The first of these relationships involves the needs of the underneath space as regards the underside surface of the floor structure. This surface is called a ceiling, and is discussed in Sec. 2.4.

The second relationship derives from the relations between the two separate spaces—above and below the floor structure. The floor/ceiling system represents a barrier between these two spaces, and must deal with the various concerns illustrated in Figure 2.2.

The third relationship involves the various building service elements that must typically be incorporated into the floor/ceiling system. This usually relates more to the lower space, but may also relate to the upper space, as shown in Figure 2.3. In some situations, floors on the ground may also need to facilitate some of these services.

Specific means for achieving floors and general floor/ceiling systems are discussed in Chapter 4 in terms of the various products and components commonly used. Common systems are discussed in general in Chapter 6, and several specific examples are illustrated in the case studies in Chapter 7.

Selection of materials and systems and the specific detailing and dimensioning of floor/ceiling systems relates strongly to building code requirements and the need for consideration of the various spatial barrier functions described in Figure 2.2. General system development must also relate very much to the need for integration of building systems. Floor/ceiling systems represent a situation where the architectural, structural, mechanical, electric, fire control, lighting, sound control, signage, and possibly other individual design concerns merge in a tight spatial context.

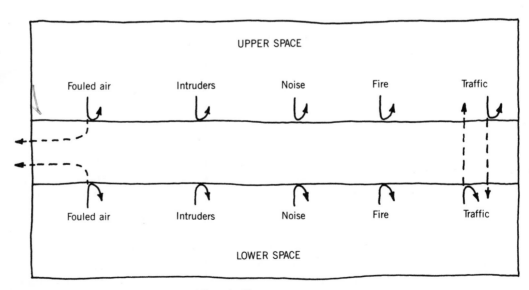

Figure 2.2   Barrier functions of floor/ceiling systems.

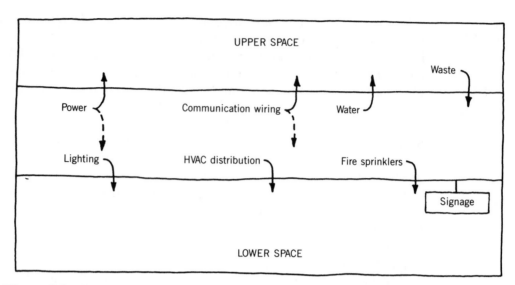

Figure 2.3   Accommodation of building services in floor/ceiling systems.

2.3 INTERIOR WALLS

Walls that form interior spaces include both fully interior walls and the inside surfaces of enclosing, exterior walls. Exterior walls are major elements in the control of the building's interior environment, functioning as barriers between the indoors and outdoors, and functioning as selective filters, as shown in Figure 2.4. The needs for these barrier and filter functions provide major criteria for the general design of the exterior walls.

Interior walls—that is, those that separate only individual interior spaces—also usually have some barrier or selective filter functions, although their general situation is different from that of the exterior walls. What is generally not critical here is the need to maintain thermal differences between the spaces on the two sides of the wall. Likewise, the need for airtightness and general weather sealing is absent. While possibly exposed to interior view, the interior walls do not constitute major elements of the building's exterior appearance. Finally, interior walls seldom contain windows—a principal feature of most exterior walls.

On the other hand, walls are often made with the same basic structures and materials, regardless of their location. All walls may need to accommodate doors, with the various requirements described in Sec. 2.5. And, in general, there are many typical requirements that are the same for all walls.

Walls must be adequate for their own support in any case, but must often also serve other structural functions, such as support of roofs or floors or bracing for wind or seismic effects on the building (as shear walls). Even when walls need only support themselves, other elements of the building structure must often be integrated into the wall construction, so as not to otherwise intrude on interior spaces. Walls that do not perform any major structural tasks, other than to support themselves, are commonly described as *nonstructural walls*. Nonstructural interior walls are generally called *partitions*, although the term *partitioning* refers in general to the defining of individual interior spaces.

Two major concerns for interior walls are their ability to function in the control of fire and sound. These properties may also be of concern for exterior walls, but do not hold quite the same level of importance in design. Major considerations for these and

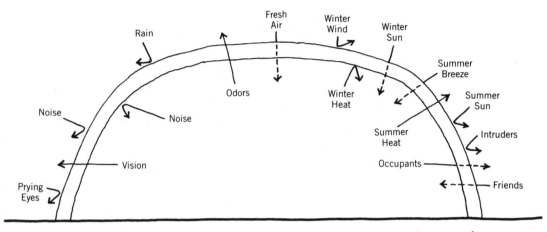

Figure 2.4   Visualization of the building enclosure system as a barrier between the indoors and outdoors, with selective filter functions.

other special functions for walls are discussed in Chapter 5. Concerns for fire are the motivation for a major portion of building code requirements.

2.4 CEILINGS

The overhead boundary of an interior space is called its ceiling. From a construction point of view, a ceiling may be a separate element, or simply be the underside of an overhead structure. Ceilings are highly visible and are usually of major concern in interior design.

Ceilings are often the location of many building service elements, such as lighting, HVAC registers, fire sprinklers, smoke detectors, and signs. The separate design of each of these elements or subsystems, as well as their total integration with the general ceiling construction, is a major design activity.

In construction work interior spaces are sometimes said to have "no ceiling," meaning that nothing special is done to create a ceiling—it is simply the unfinished, underside of the overhead structure (see Figure 2.5). More often, however, something *is* done, if nothing more than some careful cleaning or painting of the underside of the structure.

In many situations, a separate ceiling construction is developed, either attached directly to, or suspended from, the overhead structure. Thus a floor/ceiling or roof/ceiling sandwich of sorts is created, and often encloses an interstitial space that is used to contain various building service elements, such as wiring, piping, ducting, and minor equipment. The full design of this system and all of the contained elements is a major integrated design problem.

In some ways, ceiling finishes represent a major opportunity for freedom of choice of materials. There is a general lack of concern here for effects of contact wear; ceilings are not walked on and are generally out of reach. Fragile materials can be used, although there is nothing wrong with hard, durable ones either.

2.5 DOORS

Doors serve primary functions as traffic control devices, and many specific functions derive from this requirement. Doors are usually placed in walls and must also relate to

(a)

(b)

Figure 2.5   Ceiling developed as the untreated underside of the overhead construction: (a) Sitecast concrete "waffle" system. (b) Wood rib and plank deck units. (c) Exposed steel framing system.

(c)

Figure 2.5 *(continued)*

the functional requirements for the wall in terms of barrier and filter requirements. A soundproof wall with a leaky door is useless.

As shown in Figure 2.6, the basic construction elements typically consist of a rough opening in a wall, a door frame, and the operable door. In some situations, where traffic control or barrier functions are not required, the door itself may be omitted. In this case, all that is required is the hole in the wall (possibly with a frame to trim it), which is then described simply as a *doorway*. Even when a door is added, the hole plus the frame is frequently referred to as the doorway for the door.

A major distinction is made for doors that occur in the exit pathway for occupants in case of a fire. Codes require many features for these doors, including considerations for size, operating hardware, direction of swing, and so on.

Doors are potential major barriers for persons with limited facilities, such as those who are wheelchair-bound. As for fire exits, codes may require certain dimensions and special features for door operation for a truly barrier-free environment. Facilities designed specially for the aged, the very young, or persons with particular handicaps will have particular concerns. However, most buildings for public use or those where persons are employed must now be designed to some degree as barrier-free.

Doors typically have a number of hardware accessories that must be carefully chosen to satisfy all of the functional requirements for the door. This makes it necessary in the construction design process to thoroughly analyze the door requirements and to specifically identify each door in the building drawings. This is normally

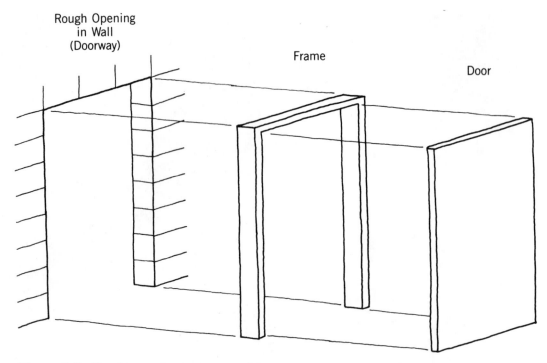

Figure 2.6 Fundamental components of doors.

done by giving each door an identifying symbol or number, which is referred to a table (called the door schedule) and/or the design specifications, where the door itself, all the hardware and other accessories, plus the details for the frame will be precisely described.

2.6 STAIRS

Buildings with floors at more than one level ordinarily require stairs. Other devices for intralevel movements may also be provided—such as ramps, elevators, or escalators—but the stair is the most primary device.

Movement in tall buildings will occur mostly with elevators, but stairs are still necessary for fire exits. In fact, the code requirements for fire stairs are usually primary sources of criteria for design of most stairs. This includes the design of the stair itself, the construction surrounding it, and the access doors to the stairs.

Planning of stairs is a major design problem (see Figure 2.7). The stairs themselves take up considerable plan space which is generally not usable for the functional activities of the building occupants. Thus tight planning is usually desirable. Placing of stairs in the general building plan is also critical to their accessibility and safe use as fire exits. While the rest of the floor plan may be essentially a two-dimensional planning problem, stairs are vertically continuous and must be planned three-dimensionally.

When a stairs is designated as a fire exit, its design is highly constrained by code requirements. These deal with the stair dimensions (width, riser, and tread), construction, stairway details, and door details. Special stairs, not designated as fire exits, may be free of some of these constraints, but must still be safe in general and subject to some limitations.

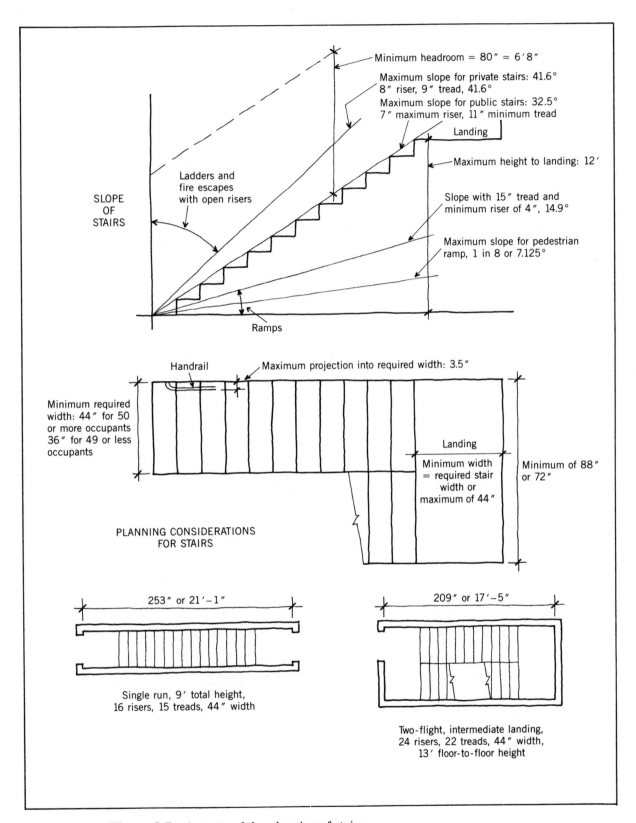

Figure 2.7 Aspects of the planning of stairs.

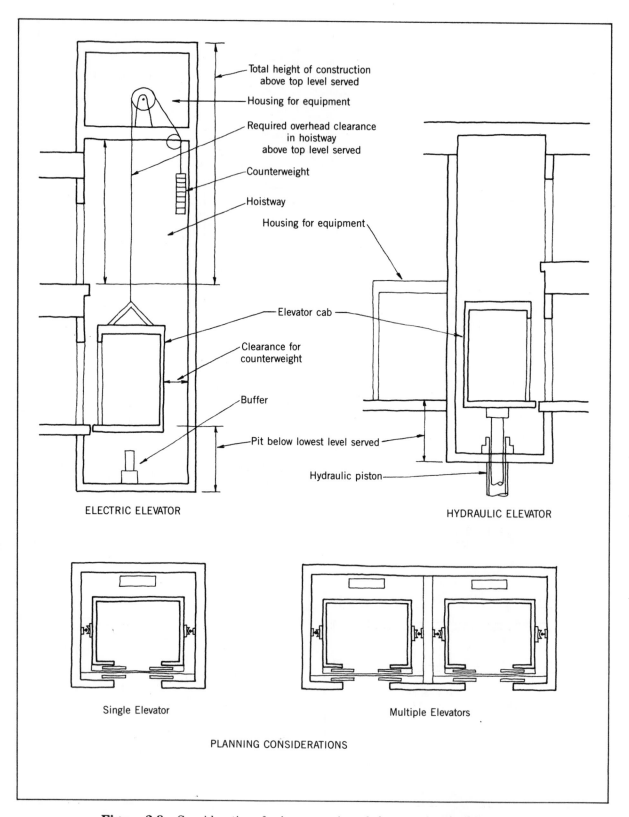

Figure 2.8 Considerations for incorporation of elevators in a building.

2.7 ELEVATORS

Elevators are now quite common in all buildings with more than one level, other than single-family houses. They may well account for the majority of intralevel movement, but do not qualify as fire exits—thus stairs must usually also be used. The stairs are also options when the elevators are not operable for any reason.

Elevators are expensive and are generally designed to minimum requirements for economy. Major concerns are for speed and the total number of users. For commercial buildings it is common to have more than a single elevator, and, of course, for very large and very tall buildings, many separate elevators.

As with stairs, in addition to their basic cost, elevators take up considerable space that is not usable for other building functions—notably those deriving income for the building owners (see Figure 2.8). Besides the plan area occupied by the elevators and the construction surrounding them, space must be provided above the highest level served and below the lowest for construction and equipment. It behooves designers with responsibility for building planning to become familiar with the ordinary space demands for elevator systems.

REFERENCES

Architectural Graphic Standards, 8th ed., Ramsey and Sleeper, New York: Wiley, 1988.

Mechanical and Electrical Equipment for Buildings, 7th ed., Stein, Reynolds, and McGuinness, New York: Wiley, 1986. chap. 23.

Sweet's Catalog Files: Products for General Building and Renovation, New York: McGraw-Hill, annually published, div. 14.

3

Materials

Materials for building construction include traditional ones long in use and new ones emerging as recent technological developments. The following discussion treats primarily the most common and widely used materials for current building construction. Emphasis here is on basic construction; finish materials for surfacing of walls, ceilings, and floors abound in great variety, but will be treated only by general categories in this book.

3.1 WOOD AND WOOD FIBER PRODUCTS

The enduring large stands of forests in North America constitute a resource for wood for many purposes. For buildings, major uses are the following (see Figure 3.1):

Solid wood, sawn and shaped: This includes structural lumber plus elements used for trim, shingles, flooring, window and door frames, and finish wall surfacing.

Glued laminated elements: These include plywood, timbers made from multiple lumber pieces, and various other products.

Fiber products: These are produced with wood fiber as a major bulk ingredient and include paper, cardboard, compressed fiber panels, cemented shredded fiber units, and cast products with wood as a principal aggregate.

Wood is organic and is subject to decay over time (by simple aging or by decomposition), to rot, and to consumption by insects. A critical aging process occurs in the early weeks and months after a tree is cut down. For solid sawn pieces, a major property is the degree of retained moisture, which generally indicates the extent to which the wood has cured from its fresh-cut (green) condition. The concern is not for the moisture itself, but for the extent to which natural shrinkage of the material has occurred, as this is what mostly causes warping, splitting, and the general development of flaws, shape changes, and dimensional changes.

For interior uses, wood is often left exposed for the attractiveness of its natural colors and grain patterns. Elements in the wood such as knots—which are actually flaws—often give the exposed wood a richer texture. However, even "natural" wood is often protected by coatings to develop resistance to wear, enhance its color, resist staining, and so on.

While wood comes basically from renewable sources, the quickest renewal (fast growth) is achieved with trees that are functional primarily as fiber sources and not for solid sawn products. Large old trees, of the types used for the best solid wood

Figure 3.1 Common construction elements of wood: solid sawn lumber, glued laminated plywood, finish woodwork, wood fiber panel elements, wood fiber paper.

products, are a dwindling source. For this, and for other reasons of economy, fiber products and some forms of laminated products are steadily replacing solid wood in many applications. Similarly, fiber products are also displacing plywood from many uses involving low structural demand.

Wood is still the structural material of choice in the United States for most ordinary construction and is mostly displaced only when there is heightened concern for fire or need for some special property that is beyond the potential of the materials. Even if it eventually recedes significantly as a solid sawn product, wood will endure as paper and other fiber-based products.

INFORMATION SOURCES

National Forest Products Association

American Institute for Timber Construction

Western Woods Products Association

American Plywood Association

3.2 METALS

Metals are used extensively in building construction, although development of new synthetic materials works continuously to replace them in many common uses. As long as their basic properties (significantly: strength, stiffness, hardness, etc.) are critical to certain tasks (notably: hardware and major structural uses), and their cost is reasonable, usage generally continues.

Steel

Steel is the most widely used metal for construction, including both size and usage ranges that are greater than that of any other material (see Figure 3.2). Hardly any form of construction can be achieved without some steel products. Nails and other connectors for wood, reinforcement for concrete and masonry, and innumerable hardware items are required in every building.

The most impressive uses of steel, however, are for the large rolled beams and columns for high-rise building frames and the stranded cables used for tension structures. Even for structural applications, however, extensive uses are for modest elements, such as formed sheet steel decking.

When exposed to air and moisture, steel rusts—a progressive condition that eventually destroys the material completely. Large, thick pieces may take a long time to rust away significantly; therefore they endure for a reasonable time. The thinner the piece, or the more crucial it is to maintain its full cross section (for structural use, for example), the more serious it becomes to mount some rust prevention method. Rust prevention is usually achieved by coating the steel surface with other metals or a rust-inhibiting paint. Steel reinforcing bars may be coated with plastic where corrosion inside masonry or concrete is a concern. Special steels can also be produced (so-called *stainless* steel or various rust-arresting steels), but these are usually quite expensive and not all steel elements can be produced with them.

Steel structures are noncombustible and are used to replace wood construction where codes require a form of noncombustible construction. The light wood frame (e.g., 2 × 4's) can be duplicated in light-gage steel (formed sheet) elements for wall

Figure 3.2   Common construction elements of steel.

framing, for example. However, steel heats rapidly and steel elements soften quickly, so some protection is required for steel structures where any significant fire rating is required.

INFORMATION SOURCES

American Institute of Steel Construction

American Iron and Steel Institute

Steel Deck Institute

Steel Joist Institute

Aluminum

Aluminum is used only sparingly for structures, as its cost is prohibitive. However, it has considerable corrosion resistance in some applications, weighs about one-third as much as steel, and can be extruded or cast into complex, sharp-cornered forms—giving it an edge for usage in various situations (see Figure 3.3).

Aluminum is used in buildings most extensively for window and door frames and for decorative trim or framing for modular ceiling systems, light fixtures, and various finish, furnishing, or service elements. Aluminum foil is used as a moisture barrier and/or reflective material in various wall and roof assemblies to enhance the barrier functions of the building enclosure.

Figure 3.3 Aluminum elements develop framing and trim for a priority modular suspended ceiling system.

INFORMATION SOURCES

The Aluminum Association

Architectural Aluminum Manufacturer's Association

Other Metals

Other metals are used for special purposes, where their properties are significant. Flashing was done extensively in the past with thin sheets of copper, lead, tin, and special alloys, but is now also extensively achieved with various plastic laminates. Decorative elements of chrome, bronze, copper, or other metals are used, but are often achieved with thin coatings over a base metal or with thin films laminated in a plastic structure.

Reflective glazing is sometimes achieved with ultrathin silver or gold films adhered to glass or, now, more often laminated into a sandwich of glass and plastic.

Many metals are still used where their functional properties are significant and no existing replacement is fully competitive. However, continued developments in materials technology steadily bring forth new materials, such as fiber-reinforced concrete and high-strength plastics, that bump metals out of general usage for building construction.

3.3 CONCRETE

Concrete is a material consisting of loose, inert particles (called aggregate) that are bound together by some infilling binder. That definition could be extended to include many mixtures, but the principal ones significant to building construction are the following:

Structural concrete: This is the material produced for sidewalks and streets as well as foundations and building frames. Aggregate is typically locally available gravel, crushed stone, sand, clam shells, and so on, selected in a range of sizes so the little pieces fill the spaces between the big ones. The objective is to have the aggregate take up as much of the volume as possible, producing a tightly packed, dense mass with very little room for the binder, which consists of portland cement and water. This basic mix can be modified in various ways by additions, called admixtures, to change the color, weather resistance, water penetration resistance, and so on. Synthetic, lighter-weight aggregates may replace the gravel to reduce the unit weight by as much as 40% or so and still produce structural resistance of significant amounts.

Insulating concrete: This is ultralightweight concrete, produced with portland cement and some very lightweight aggregate, such as wood fiber or some low-density mineral substance, such as vermiculite or perlite. The mix is lightly foamed with as much as 20% air, and the unit density can be as low as 30 pcf, compared with 110 to 150 pcf for structural concrete. This material may be used for fire protection of steel elements, but is mostly used as fill on roof decks.

Fiber concrete: Research and experimentation have resulted in the use of various forms of concrete with fiber materials in the mix. Their general purpose is to add some additional tensile capacity to the concrete, reducing or eliminating the need for the usual steel reinforcement used to alter the tension-weak character of the concrete. Roof tiles, siding, and pavements are some of the current applications for this form of concrete.

Gypsum concrete: Gypsum can be substituted for portland cement in plaster or concrete. A lightweight roof deck material is produced with gypsum and an aggregate of wood chips. This is an insulative material with significant strength and stiffness for roof structural applications in controlled situations. Priority systems consist of combinations of gypsum concrete cast over modular metal supports and forming units, the forming units developing finished ceilings in some applications.

Bituminous concrete: Substitute a hot bitumen (coal tar, etc.) for the portland cement and water and you have asphalt pavement. Not much used for buildings, it is nevertheless a significant material for site development.

In most applications, structural or not, some compensation must be made to overcome the tension-weak character of the concrete. Steel reinforcement, in the form of rods or welded wire mesh, is the usual solution. The concrete encasement needs to protect the steel to prevent rusting or insulate it from fire. Other means for improving tension resistance include the use of fiber materials (as mentioned previously) and prestressing, which consists of inducing compression in structural concrete members before the occurrence of loading that produces tension. Prestressing is usually induced by high-strength steel strands stretched inside the concrete and anchored or bonded to the concrete. Maintaining the stretch of the steel develops compression in the concrete.

Concrete must be cast in forms and cured over some period of time after it is mixed and placed in the forms. Hardening results from a chemical action between the water and cement, and takes some time to occur in ordinary situations. The cast concrete must be kept moist and at a reasonable temperature during the curing period.

Mixing, transporting, casting, finishing, curing, and reinforcing or prestressing of concrete represent major cost factors. Reducing the cost of these items is often more significant than attempts to make the concrete mix itself cheaper. It is better to go for the best mix and find ways to reduce other costs—particularly the time-consuming and labor-intensive ones.

Concrete is generally the most durable material for ground-contacting construction, and is thus extensively used for foundations, basement walls and floors, pavements, and other site construction. It is also generally weather resistive and can function reasonably in an exposed condition as a raw, unprotected material. Finally, it is considerably fire resistive. All of this frees it from the need for the many concerns for protection often required for wood and steel.

Because of its functional capability of being exposed—on both the exterior and interior—concrete surfaces are frequently in view, mostly for walls and the underside of spanning roof and floor structures (see Figure 3.4). Where this is the case, and appearance is of some concern, it is necessary for the designer to deal with the various factors that influence the development of surface form and texture. These factors are discussed at length in the following chapters.

INFORMATION SOURCES

American Concrete Institute

Portland Cement Association

Prestressed Concrete Institute

3.4 MASONRY

The term *masonry* covers a range of construction that overlaps the areas of tiling, cladding, and precast concrete. Traditional forms of brick and stone masonry can still be obtained, but other processes and materials have largely displaced them. Most structural masonry for bearing walls, shear walls, or foundation walls is now produced with units of precast concrete (concrete blocks—called CMU for concrete masonry unit). What appears to be a brick wall these days is often a frame or CMU structure with thin "brick" tiles adhesively bonded to a backup material and mortar joints simulated with a filler material.

Stone is mostly used decoratively these days. Fieldstone is developed as either veneer or facing on other structures. Cut stone (granite, marble, etc.) is cut in increasingly thin slices to form facings over structures of other materials.

In the face of these trends, the crafts of stonemasonry and bricklaying slowly decline and become ever less obtainable. The enduring popularity of the form of construction (or at least its appearance) keeps it alive—but increasingly as illusion or metaphor rather than as fact.

Classic masonry consists of some inert units bonded into a contiguous mass with a binder between the units (see Figure 3.5). The binder is usually mortar—similar in

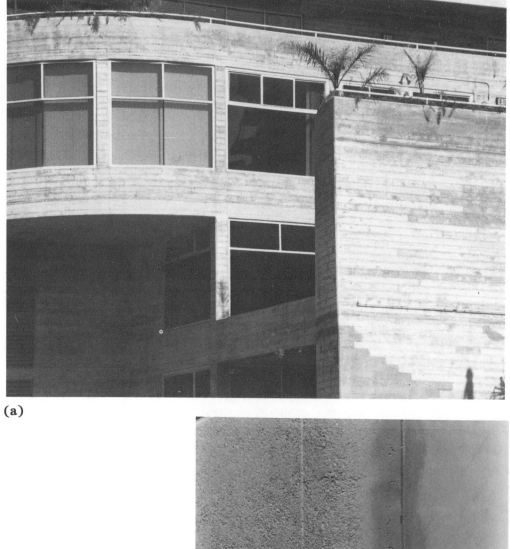

(a)

(b)

Figure 3.4 Development of various finishes for cast concrete elements: (a) Formed finish. (b) Abrasively-finished surface; exposing the stone aggregate. (c) Tooled finish, worked on the hardened concrete surface.

(c)

Figure 3.4   (*continued*)

form to that used in ancient times, but quite different today in its ingredients and properties. Units may be stone—cut or unprocessed—bricks of formed and fired clay, precast concrete (mostly as hollow blocks), or other elements, such as hollow tiles of gypsum or fired clay.

It is desirable to have masonry consist mostly of the units, with a minimum of mortar in thin joints between the units. It is best to rely on the mortar only for compressive resistance, although it is desirable to have it adhere well to the units to bond them together for general stability of the construction.

Overall, the strength and stability of masonry construction depends on the units, their arrangement, the mortar joints, and—above all—on the craft of the workers who prepare the units, mix the mortar, lay up the units, and protect the construction while the mortar hardens. When it all works, it is great, but in modern times the biggest drawback to the use of classic masonry construction is its dependency on the craft of the workers.

Masonry units must be closely fit together, so the size and shape of the units, the patterns of their arrangement, the general form of the construction they are forming, and the required edges, corners, openings, intersections, supports, and other form-related concerns must be carefully developed. This is especially critical for work exposed to view. It is also most critical for CMU construction; bricks and stones can be cut to trimmed sizes and shapes, but concrete blocks must be used in whole-piece form. Special CMU shapes and sizes are provided, but some dimensional and modular control is necessary.

Like concrete, masonry is tension-weak, which needs consideration in structural applications. The strongest masonry structures are usually those developed as *reinforced masonry,* which uses steel rods in a manner similar to that for reinforced concrete. This is of greatly heightened concern for structures required to resist major earthquakes or violent windstorms.

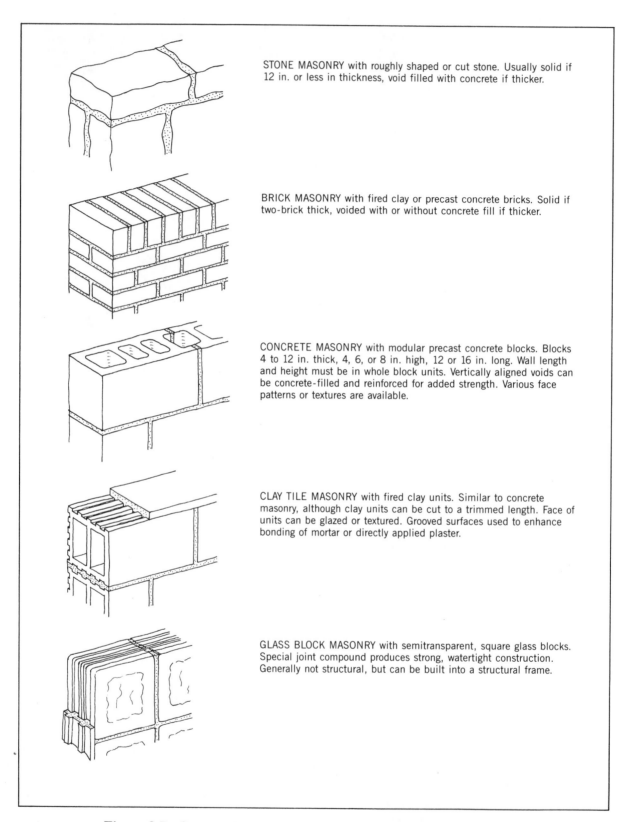

STONE MASONRY with roughly shaped or cut stone. Usually solid if 12 in. or less in thickness, void filled with concrete if thicker.

BRICK MASONRY with fired clay or precast concrete bricks. Solid if two-brick thick, voided with or without concrete fill if thicker.

CONCRETE MASONRY with modular precast concrete blocks. Blocks 4 to 12 in. thick, 4, 6, or 8 in. high, 12 or 16 in. long. Wall length and height must be in whole block units. Vertically aligned voids can be concrete-filled and reinforced for added strength. Various face patterns or textures are available.

CLAY TILE MASONRY with fired clay units. Similar to concrete masonry, although clay units can be cut to a trimmed length. Face of units can be glazed or textured. Grooved surfaces used to enhance bonding of mortar or directly applied plaster.

GLASS BLOCK MASONRY with semitransparent, square glass blocks. Special joint compound produces strong, watertight construction. Generally not structural, but can be built into a structural frame.

Figure 3.5   Common forms of masonry for building interiors.

INFORMATION SOURCES

Brick Institute of America

Clay Products Association

National Concrete Masonry Association

3.5 PLASTICS

Plastics are used extensively in buildings as solid formed elements for trim and hardware, as film for moisture barriers or in laminates, as foamed insulation, and for many forms of coatings. As in its early days, plastic is often used to imitate other materials. (Is it real or is it *plastic?*) Thus wood paneling is often a photograph of wood embedded or laminated into plastic and mounted on a compressed wood fiber panel; wood strip siding is duplicated in form and appearance with formed vinyl sheets; metal trim or surfacing is actually thin metal film over plastic; brick, Spanish clay tile, mosaic tile, marble, and just about anything can be imitated with plastic in Disneyland style.

Basic chemistry, forming methods, and combinations of plastic with other materials can be varied extensively to respond to many demands. Potential problems include concerns for fire resistance, low strength or stiffness, wear resistance, stability over time, and possible toxic effects.

Many paints, varnishes, and other coatings have a plastic base. Multiple coatings may produce combined effects, with each coating adding something additional to the total effect.

INFORMATION SOURCE

Society of the Plastics Industry

3.6 MISCELLANEOUS MATERIALS

Many materials can be used for various purposes for building construction. Some are used in unique applications; some are used as substitutes for more traditional materials—for cost savings or for some enhanced property that is superior to the material it is replacing.

Paper and Other Fiber Products

Paper is used with various elements of building construction. Two common uses are the facings for gypsum drywall units (a sandwich with a gypsum plaster core) and the backup material for stucco. Paper can be coated—with plastic, but also with aluminum foil, wax, or bitumen. It can also be laminated; for example, with glass fiber strands for reinforcement.

Cardboard in solid form is more or less thick paper, intermediate between ordinary paper and various forms of compressed wood fiber panels. Many variations, including ones with coatings or lamination, are possible. Concrete forming, such as that for round columns, is sometimes accomplished with heavy cardboard elements.

Corrugated cardboard is a sandwich of alternating layers of flat and pleated or corrugated paper, resulting in a material with considerable strength and stiffness but relatively light weight. As with other paper products, special coatings and reinforcements are possible.

Composite Materials

Composite elements are those in which two or more different materials are blended in a way such that the materials retain their individual identity but share some effort with the other material(s) with which they are blended. Examples are reinforced concrete (concrete plus steel) and laminated glazing (glass plus plastic). Increasing use is being made of composites in building construction. In some cases these produce entirely new elements with many new potential possibilities. In other cases they simply permit an extended usage of old familiar materials beyond their previous limitations.

4

Elements

General concerns for basic elements of interior construction are discussed in Chapter 2. This chapter presents materials relating to the specific use of products and systems for interior construction. Special concerns for various properties of the construction are discussed in Chapter 5. Systems for achieving whole buildings are discussed in general in Chapter 6 and illustrated for particular cases in the examples in Chapter 7.

4.1 FLOOR STRUCTURE

General problems of floors and floor/ceiling systems are discussed in Chapter 2, Sections 2.2 and 2.4. The following material presents information on the two basic types of floor structures—pavement slabs and framed systems with decks.

4.1.1 Pavement Slabs

The usual means of achieving a working surface for sidewalks, driveways, basement floors, and floors for buildings without basements is to cast concrete directly over some prepared base on top of the soil. As shown in Figure 4.1, the typical construction for this consists of the following components:

1. A prepared soil surface, graded to the desired level; compacted, if necessary to avoid subsidence.
2. A coarse-grained pavement base, usually predominantly fine gravel and coarse sand; also compacted to some degree, if more than a few inches thick. Three or four inches is common for floor slabs.
3. A membrane of reinforced, water-resistant paper or thick plastic film (6 mil minimum), used where moisture intrusion is especially severe.
4. The concrete slab.
5. Steel reinforcement; often of heavy-gage wire mesh.

While the basic construction process is quite simple, there are various considerations that must be addressed in the development of details and specifications for the construction. The following discussion pertains mostly to simple floor slabs for building interiors, where weather exposure and exceptionally heavy vehicular traffic are not of concern.

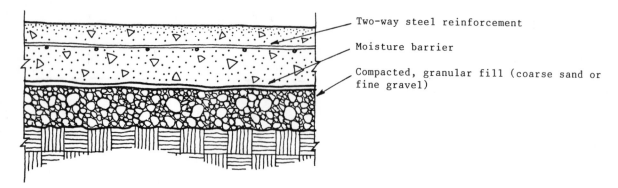

Two-way steel reinforcement

Moisture barrier

Compacted, granular fill (coarse sand or fine gravel)

Figure 4.1   Common form of concrete pavement for a building floor.

Thickness of Pavements

For residences and other situations with light traffic, a common thickness is a nominal 4-in. slab; actually 3.5 in., if standard lumber 2 × 4's are used as edge forms. With proper reinforcement and good concrete, this is an adequate slab for most purposes.

Where some vehicular traffic is anticipated, or where other heavy concentrated loads—such as those from tall, heavy partitions—are expected, it may be desirable to jump to the next logical thickness of 5.5 in., based on using 2 × 6's for edge forms. At this thickness, it may be possible to avoid providing individual wall footings for nonstructural partitions, a simplification in construction probably justifying the cost of the additional concrete for the slabs.

For very heavy trucks, for storage warehouses, or for other situations involving heavy concentrated loads, thicker pavements may be necessary. However, the nominal 4-in. and 6-in. slabs account for most building floors.

Reinforcing

For 4-in. slabs, reinforcing is usually accomplished with welded wire mesh. For thicker slabs—or as an alternative for 4-in. slabs—small-diameter rebars may be used with somewhat wider spacings than those in the wire mesh. Even when mesh is used, some extra rebars may be provided at edges, around openings, or at other critical locations.

The objective for the reinforcement is to reduce cracking of the concrete, due primarily to shrinkage during curing, differential volume changes due to temperature changes, and some unequal settlement of the pavement base. Cracking is especially undesirable in the top (exposed surface) of the slabs, so the reinforcement should be held up during casting to be relatively close to the top surface.

Another means for reinforcing—or basically altering—the concrete is to add fiber materials to the mix, a growing practice for all exposed slabs. In cases where considerable movement of the base is expected, pavements may need to be prestressed or developed as framed systems on grade. Figure 4.2 shows two possible forms for the creating of strong ribs in a cast slab, for use as wall strip footings or as parts of a framed system on grade.

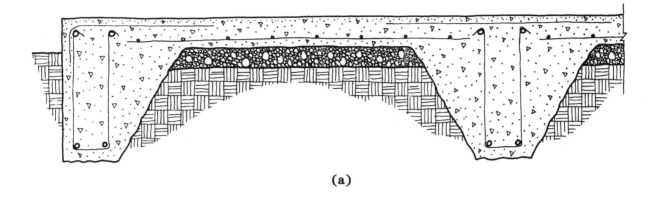

(a)

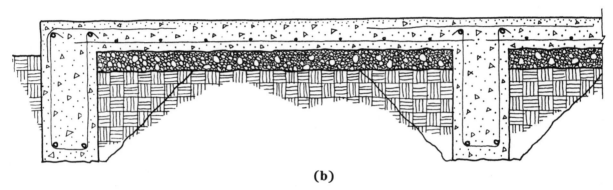

(b)

Figure 4.2 Common forms for concrete framing cast directly on the ground: (a) Beams trenched and cast with slab. (b) Beams separately formed and cast.

Joints

It is generally desirable to pour paving slabs in small units, primarily to control shrinkage cracking. This results in construction joints, which may not relate well to development of floor finish materials. Where permanent wall construction is established, joints should be located at these points rather than in the middle of floor spaces.

Larger pours can be made if control joints are formed or cut in the slab. Movements at these joints may be small for interior floors, but they should still be located at points that will not cause problems with floor finishes if possible.

Surface Treatment

If the top of a paving slab is to be exposed to wear, it should be specially formed for this purpose during casting of the concrete. A hard, smooth surface is generally desired, typically developed with fine troweling with a steel trowel and possibly the addition of some hardening materials to the finished surface.

For a slip-resistant surface, intended to reduce accidents when the floor is wet, a grit material may be added to the surface. Smoothness is also generally not highly desirable when other finish materials are to be used, especially ones that must be attached with adhesives. In the latter case, the steel troweling may be omitted and a rough leveling of the surface acceptable.

Framed Construction on Grade

As previously described, it may sometimes be necessary to provide an actual framed construction on top of the ground. Two situations that often make this a possible necessity are the following:

1. When finished grade must be achieved with a considerable amount of fill, making it likely that major settlement of the pavement base will occur.
2. When deep foundations are used to support columns and walls with very little settlement. Even though the paving base may be good, some relative settlement is likely.

Framed systems used in this case are usually variations on the typical slab-and-beam systems, described in the next section. For modest spans, beams may be created by simple trenching, as shown in Figure 4.2a. For longer spans, beams of considerable depth may be needed and a separate forming of beams required, as shown in Figure 4.2b.

4.1.2 Framed Floors

When another space exists below a floor, a framed structure must be used to span some clear distance to achieve a platform for the floor (see Figure 4.3). Various

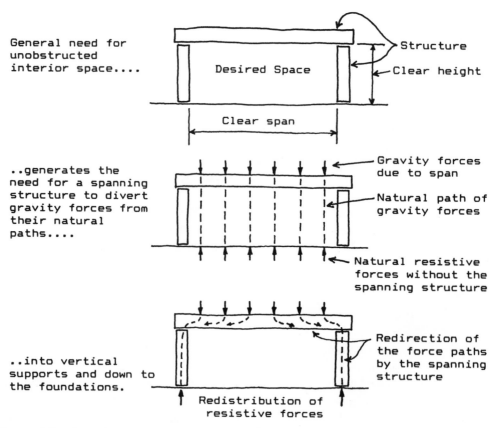

Figure 4.3 Considerations for framed roof and floor systems with regard to generation of interior spaces.

materials and systems can be used to achieve the spanning structure, with choices relating to many considerations, including the following:

1. The materials and basic construction of the rest of the building
2. The dimension of required horizontal, clear spans
3. The form and layout of supports for the floor structure: bearing walls, columns, widely spaced piers, and so on
4. Penetrations of the basic floor structure for stairs, elevators, duct shafts, or other elements
5. Any special requirements for loads: exceptionally heavy loads (such as in warehouses or libraries), concentrated loads (wheel loads, equipment, etc.), or hanging loads
6. Special considerations for floor finishes or supported ceiling construction

Many special systems can be used, but the most common systems are variations on simple deck-and-beam systems, with a deck of some flat material supported by regularly spaced beams. Most floor structures have relatively modest spans permitting a wide range of spanning elements, typically of wood, steel, or reinforced concrete.

Figure 4.4a shows a common system in light wood frame construction, consisting of a structural plywood deck and very closely spaced wood joists. This is reasonable for spans up to about 20 ft. Beyond this span, other elements may be substituted for the solid lumber joists; one example is the composite plywood and timber I-beams shown in Figures 4.4b and c.

Common in the past, but less widely used today, is the so called *heavy timber* system that uses a nominal 2-in.-thick plank decking and larger timber beams at wider spacing (see Figure 4.4d). (Ordinary wood joists are typically 12, 16, or 24 in. on center.) The exposed heavy timber structure offers the advantage of some fire resistance, while the light wood frame must be covered to obtain any rating.

Although the exposed, heavy timber frame was common in the past for both utilitarian (barns, etc.) and higher-quality buildings (churches, etc.), the construction is now usually premium priced and the craft for its execution not so easy to obtain.

As is frequently mentioned here, the light wood frame is the definitive choice for the structure of buildings in the United States, except when circumstances prevent its use. Other choices must be compared with the light wood frame and justified on the basis of some extenuating circumstance, a major one being code requirements for fire ratings.

The general form of the light wood frame deck-and-joist system can be developed with steel elements, as shown in Figure 4.5a, in this case using steel open-web joists. Analogous to the heavy timber system is the system shown in Figure 4.5b, using a slightly heavier deck and heavier beams (rolled W sections here).

While wood framing is used almost exclusively to support plywood, wood plank, or wood fiber decks, steel framing can be used to support almost any deck system, including plywood, sitecast concrete, or precast concrete, as well as a wide range of steel decking. For wood and steel decks it is common to use a concrete fill on top of the deck for many reasons, such as the following:

1. It adds a fire barrier/insulation on top of the floor system.
2. It creates a smooth surface, especially on formed steel decks.

(a)

(b)

Figure 4.4   Wood systems for floors: (a) Ordinary joist system with plywood deck and solid sawn joists of structural lumber. (b) Composite elements of plywood and solid sawn wood. (c) Heavy timber system with glued laminated beams.

(c)

Figure 4.4   (*continued*)

3. It adds stiffness and a feeling of solidity to the deck.
4. It improves the acoustic separation character of the floor system.

In the past, it was common to use the existence of the concrete fill as an opportunity to bury wiring systems for power and communication. This is still done, although other options exist for many situations, most of which allow more ease of modification as needs rapidly change in many commercial occupancies.

Concrete floor systems may be sitecast or precast. As mentioned previously, concrete decks may be used with steel framing. In fact, concrete will ordinarily be used in one way or another with any steel floor system (see Figure 4.6). This may occur in any of the following ways:

1. As inert concrete fill on top of a structural deck of formed sheet steel; the steel deck carries the concrete, as well as other loads
2. As a composite deck system, with the concrete fill and the formed steel deck interlocked to work together as a single structural unit
3. As a structural concrete deck, using a formed steel deck strictly for the casting of the concrete
4. As a structural sitecast concrete deck poured on removable forming
5. As a precast concrete deck, with units supported by the steel framing

Each of these choices has some circumstances that make it a more feasible alternative; however, they are also somewhat competitive for similar situations, with choices made as much on the basis of personal preferences as any other reason.

Wholly sitecast concrete systems include the common options shown in Figure 4.7. The wood or steel deck and beam system can be obtained in cast concrete (Figure 4.7a), and is indeed probably the most common system, due to its great adaptability. Other systems may be justified or preferred in various situations. The flat slab and flat plate systems (Figures 4.7b and c) offer a clean underside form, which may be

(a)

(b)

Figure 4.5   Steel floor framing systems: (a) System with steel trusses and formed sheet steel deck; trusses develop both the joist system and the column-line supports. (b) System with rolled steel W sections and longer-spanning steel deck.

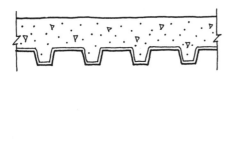

STRUCTURAL STEEL DECK/CONCRETE FILL

Usually structural grade, lightweight concrete. Concrete may be inert or act in composite action with the steel deck.

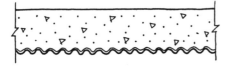

STEEL DECK FORM/STRUCTURAL CONCRETE SLAB

Basically a reinforced concrete slab. Steel form deck may act in composite action as the bottom reinforcement for the slab.

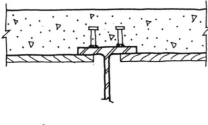

CONCRETE SLAB/COMPOSITE ACTION WITH BEAMS

Steel shear developers welded to top of steel beams. Removable wood forming.

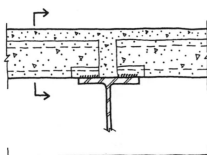

PRECAST CONCRETE DECK UNITS WITH SITECAST CONCRETE TOPPING

Metal connectors cast into deck units and site welded to steel beams.
Adjacent deck units attached by sitecast concrete joint.
Structural concrete topping bonds with precast units to form composite structural action.
Hollow tubular voids in deck units can be used for wiring or piping, but not often done.

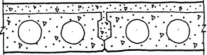

Figure 4.6   Concrete floor decks with steel framing.

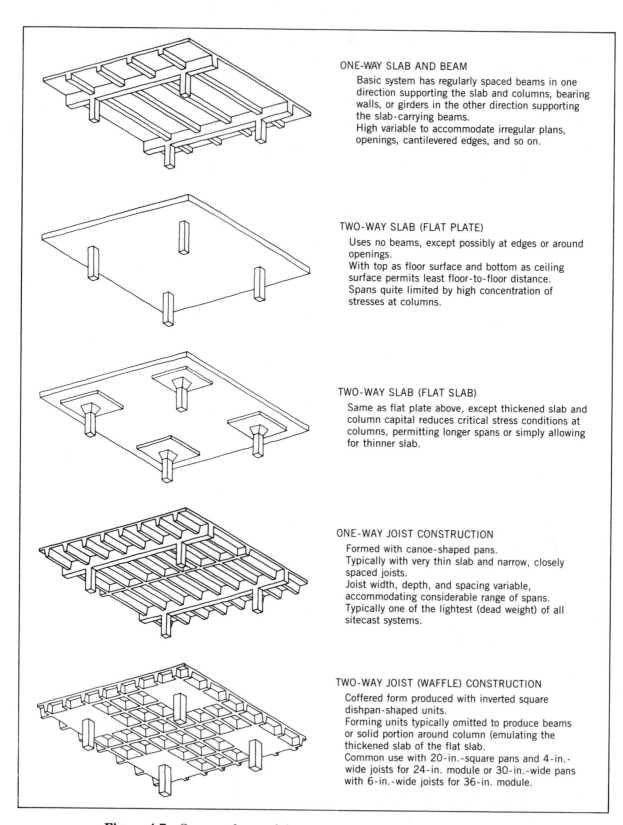

ONE-WAY SLAB AND BEAM

Basic system has regularly spaced beams in one direction supporting the slab and columns, bearing walls, or girders in the other direction supporting the slab-carrying beams.

High variable to accommodate irregular plans, openings, cantilevered edges, and so on.

TWO-WAY SLAB (FLAT PLATE)

Uses no beams, except possibly at edges or around openings.

With top as floor surface and bottom as ceiling surface permits least floor-to-floor distance.

Spans quite limited by high concentration of stresses at columns.

TWO-WAY SLAB (FLAT SLAB)

Same as flat plate above, except thickened slab and column capital reduces critical stress conditions at columns, permitting longer spans or simply allowing for thinner slab.

ONE-WAY JOIST CONSTRUCTION

Formed with canoe-shaped pans.

Typically with very thin slab and narrow, closely spaced joists.

Joist width, depth, and spacing variable, accommodating considerable range of spans.

Typically one of the lightest (dead weight) of all sitecast systems.

TWO-WAY JOIST (WAFFLE) CONSTRUCTION

Coffered form produced with inverted square dishpan-shaped units.

Forming units typically omitted to produce beams or solid portion around column (emulating the thickened slab of the flat slab.

Common use with 20-in.-square pans and 4-in.-wide joists for 24-in. module or 30-in.-wide pans with 6-in.-wide joists for 36-in. module.

Figure 4.7   Common forms of sitecast concrete floor systems.

functionally useful in obtaining overhead clearances in the space below the structure, or simply be cleaner in appearance where this is a factor.

Narrow ribs and thin decks can be used to produce relatively light, longer-spanning systems in sitecast concrete. With ribs in one direction only (Figure 4.7d) this is called concrete joist construction. A unique possibility in the cast concrete material is to create a two-way, intersecting rib system (Figure 4.7e), called a waffle system, as its underside resembles a huge waffle. The ribbed systems are ordinarily produced with some priority forming units, typically available in a modular set for forming the basic system and special edges, openings, and so on.

An advantage of concrete systems is the possibility of using them in an exposed condition while still retaining a high fire rating. This is less so with the ribbed systems, due to the relative thinness of the elements and generally reduced concrete cover (fire insulation) for the steel reinforcement. While it is popular for its rich texture, the waffle system is quite limited for use as a floor structure by this qualification.

Precast concrete systems also abound in form, although a few very common elements are produced quite widely. Some of the most common spanning elements are those shown in Figure 4.8. These are used mostly for singular spanning situations and can be supported by other precast elements or by ordinary concrete, steel, or masonry structures.

In general, concrete spanning systems are quite expensive and are used only when their various properties are of significant value. Major factors in this regard are general weather and exposure resistance, resistance to wear, natural fire resistance, ability to be exposed to view and retain fire ratings, and natural blending with other construction of concrete or masonry.

4.1.3 Floor Supports

Framed floors are ordinarily supported by some combination of walls and columns. Most framing systems are one-way spanning in nature and need a secondary spanning support system when used for multiple spaces. The framing systems and support systems must be integrated in various ways; some considerations for this are as follows:

1. The spanning character of the framing is established by the support system, which effects the span dimensions, one or two-way spanning situation, and any repeating modules in the framing layout.
2. Support elements are necessarily permanent construction and are important fixed elements in the building plan.
3. For simplest structural planning, support elements should be vertically aligned in multistory construction.
4. Lateral bracing is usually achieved with vertical elements at the location of

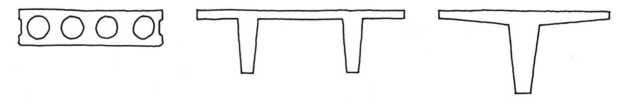

Figure 4.8 Common forms of precast concrete elements used for floor systems.

supports for the framing system; therefore the framing support and lateral bracing functions must be planned for simultaneously.

Bearing walls can be achieved with various forms of construction, the most common being the following (see Figure 4.9):

1. Stud framing—mostly in light wood (2 × 4, etc.) and mostly for support of light wood, joist, or rafter framing.
2. Structural masonry—mostly with CMUs (concrete blocks); used to support just about any kind of framing: wood, steel, or concrete.
3. Sitecast concrete—used mostly with other concrete construction, but can support any framing. Foundation and basement walls represent the most frequent use.
4. Precast concrete—used mostly for exterior walls, mostly with tilt-up construction.

Bearing walls represent highly fixed, generally solid plan elements that require serious architectural planning. Some openings are possible, but the basic form must

(a)

Figure 4.9   Common forms for structural bearing walls: (a) Closely-spaced columns form a stud framing system. (b) Structural masonry; shown with precast concrete units (CMUs). (c) Precast concrete.

(b)

(c)

Figure 4.9   (*continued*)

be that of solid wall construction. If the need exists for considerable open plan space or for future rearrangement of walls, the use of bearing walls is probably questionable.

Columns may be arranged as an independent system, or may be developed in conjunction with a wall system. Small columns can be buried inside stud walls to eliminate any plan intrusion. (Developed as multiple studs or as 4×4, 4×6, etc., inside a 2×4 stud wall.) Pilasters can be used with masonry or concrete walls, representing lumps in the wall form, but at least not a whole freestanding presence in a plan.

Columns are locations of concentrated loads, and the framing that delivers the loads must be acknowledged in planning. Vertical wiring, piping, ducting, and any openings in the floor system must be coordinated so as not to conflict with locations of major framing elements.

Columns also need major foundations, which must be considered in planning elevator pits, under-floor drains, and other elements. A column may fit into the plan at a floor level, but the whole vertical structure should be considered. In tall buildings, column sizes will probably change from top to bottom of the structure, and planning at all levels should be considered. For example, an elevator shaft must be constant in dimensions throughout its height, even though bearing walls or columns at its edges change in successive levels.

The pattern of supports establishes the nature of spanning required by the framing system supported. One-way span systems need support only on two sides; continuous framing needs repetitive, reasonably equally spaced supports; two-way spanning systems need special supports that permit the appropriate two-way action.

Planning of support systems must be done as a coordinated effort with general architectural planning development. Architectural planning may lead structural planning and the logical framing system may emerge. On the other hand, if an early decision is made to use a particular type of structure—for example, when exposed construction is used—the needs and limitations of the structure must be dealt with in developing architectural plans.

4.2 FLOOR FINISHES

Most floor structures receive some form of finish as a wearing surface. This is more or less functionally required, depending on the materials of the construction of the structural deck. Concrete slabs can be finished to a high-quality surface condition, if so desired. On the other hand, structural plywood decking is usually quite rough, and gets rougher as nailing of edges is achieved.

Finishes abound, but a few common ones account for most floors. The following discussions deal with common situations, with primary attention given to construction concerns.

4.2.1 Concrete

Concrete top surfaces, particularly of pavements or structural decks, can be brought to a finished surface condition in a variety of ways. Of primary concern usually is the simple truing up or smoothing out of the surface and the development of as flat a surface as possible. It is also usually essential to achieve a good, dense concrete surface before the concrete attains its initial hardening—called the set.

If an applied surface, or simply additional construction, is to be placed on top of a concrete surface, the truing up may be done with the accomplishing of a relatively rough and slightly porous surface. In many situations, a thin additional layer of lean concrete fill is used to achieve a finer, flatter, generally truer surface for a finished surface or to receive thin surfacing materials, such as linoleum or vinyl tiles.

When fill is placed on top of a structural concrete slab, it may be used in a manner similar to that with fill on a steel or wood deck, for example, to contain buried wiring systems. This is usually a simpler way to contain wiring, rather than within the structural concrete. It also allows for a total rewiring in the future without disturbing the basic structure.

Concrete surfaces left exposed can be finished in various ways, depending on usage and maintenance. The concrete itself may be made as smooth as possible, or it may be deliberately roughened to reduce slipperiness when wet or oily. Grit of some form may also be embedded to achieve the nonslip surface. On the other hand, paint, sealers, or wax may be used to enhance the smoothness of the surface.

For construction detailing and building dimensions, it is important to anticipate the desired finishes of floors, particularly when the surface of the structural concrete must be depressed to permit additional construction for finishes.

INFORMATION SOURCE

Portland Cement Association

4.2.2 Wood

Wood flooring is still widely used, although other floor finish materials have displaced it in many situations. Hardwood flooring used to be common for residential occupancies, but is now rare, since carpeting is now so widely used.

Wood flooring is attractive, and is now mostly reserved for feature usage—as with marble, ceramic tile, and so on—except for some specific uses, such as gymnasiums. Wood flooring is now mostly installed in prefinished blocks or strip panels and is often attached with adhesive materials, rather than nailed down.

INFORMATION SOURCE

National Oak Flooring Manufacturers Association

4.2.3 Carpet

Wall-to-wall carpet has emerged as a universal floor covering. Tough, stain-resistant synthetic fibers and mass production, as well as simplified installation, have made this a diverse, easily installed, and easily replaced covering material. Reduction in the number of smokers has also made it more feasible, burns being the nemesis of synthetic fibers.

Carpet with reasonable padding is also fatigue reducing for occupants who must stand on the floor for many hours. In general, it is usually a pleasant and highly popular floor surfacing for building users.

Carpet must be applied over a reasonably smooth surface, or else it will eventually tend to reflect the surface flaws beneath it. Smoothing elements are thus often used, such as concrete fill or a layer of wood fiber board. This was not a problem when floors were routinely finished with hardwood flooring or tile, but floors are now often left with rough, structural finishes in anticipation of a fully developed finishing of some kind.

There is a great range of quality and type of carpeting, permitting considerable variety within a simple basic construction provision.

INFORMATION SOURCE

Carpet and Rug Institute

4.2.4 Thin Tile

Thin tile—literally dimensionally thin, although installation may require additional materials—may be of various types, including the following common ones:

1. Linoleum or square units of vinyl or other flexible materials
2. Thin-set ceramic tiles, attached with adhesive materials
3. Thin wood blocks or strips, attached with adhesives
4. Relatively thin units of stone, fiber-reinforced concrete, and other priority materials, attached with adhesives to emulate classic stone flooring

As with carpet, the supporting surface must be quite smooth and flat, as bumps and cracks will eventually be reflected. Thus, while the finish materials themselves are thin, necessary fillers may add to the total construction that must be provided for by the supporting structure. Linoleum placed directly over plywood will eventually reflect not only the plywood joints and nails, but even the wood grain of the face plies of the plywood.

Most hard floor finishes—of wood, stone, and ceramic tiles—can now be produced with less dimensional allowance in the construction. Not quite as thin as a coat of paint, but mostly a lot less than was required in the old days.

INFORMATION SOURCE

Tile Council of America

4.2.5 Thick Tile

Some forms of finish, essentially executed in the classic manner, may still require considerable allowance for the thickness of the complete work. Fieldstone, thick ceramic tiles, precast concrete tiles, or other elements may be placed in a mortar bed in the classic method. Depending on the size of the units, this may require several inches or as little as 1.5 to 2 in. for development on top of a structural concrete surface.

Terrazzo is essentially—in its present form—a concrete finish material, developed by a special concrete mixed with coloring agents and typically marble or other stone chips. When hardened, the surface is ground and polished to a very smooth finish. Brass strips are commonly used to separate zones of different colors to produce large floor patterns.

Terrazzo is a very functional, durable finish, although quite expensive when installed in the classic manner, with a buildup of as much as 2.5 in. of total thickness, including the final, thin layer of finish material. Many alternative techniques can also be used to simplify an installation, including use of thin-set, precast units. The good stuff, however, still takes a lot of time, a lot of money, and considerable thickness for construction.

INFORMATION SOURCE

Tile Council of America

4.3 WALL STRUCTURE

Walls bounding interior spaces occur as one of two situations: the interior walls, with interior spaces on both sides of the wall, and the interior face of exterior walls forming the building enclosure. The discussion in this section deals primarily with interior walls, although many construction issues—especially as regards finish materials—are essentially the same for the insides of the exterior walls.

Interior walls may be structural, serving to support other construction or to brace the building for lateral loads. They may also be nonstructural, serving only to support themselves (as partitions) and possibly not even full height within a building story. Even nonstructural walls, however, occur frequently at the location of columns or bracing elements such as X-bracing or trussed diagonal bracing. In any event, planning of walls must usually be closely coordinated with the general planning of the building structure.

Walls are frequently used to contain wiring, piping, or ducts, as well as various items for building services, such as outlets, switches, recessed cabinets, and so on. The planning of building service systems and detailing of installations must be done with the knowledge of the form of interior wall construction.

4.3.1 Framed Walls

The most common framed wall is the light wood frame with 2×4 studs (Figure 4.9a). The 2×4 can usually be used for walls up to about 14 ft in height, and taller walls can be framed in a similar fashion with 2×6 or 2×8 studs, accounting for most walls in ordinary construction. The wider studs also create a larger hollow (interstitial) space, which is now frequently used in exterior walls to permit additional thermal insulation.

Framed walls must be surfaced in most cases. Studs are typically quite close, so relatively thin materials can be used for this for ordinary situations. The most common surfacing for interior walls is gypsum drywall in panel form, consisting of a sandwich of two heavy sheets of paper and a precast gypsum plaster core.

Other surfacing materials commonly used are panels consisting of plywood, processed wood fibers, or composites of mineral fibers with glass fiber reinforcement. Plaster (wet wall versus dry) can be used to form high-quality surfaces, but is frequently simply applied over a drywall base. A really hard, dense surface can be achieved with portland cement plaster.

Framed bearing walls are essentially clusters of closely spaced columns (the studs). Surfacing is generally not considered to contribute to vertical load resistance, although the thin studs are braced on their weak axes by the surfacing. For shear walls, on the other hand, the surfacing is the basic shear-resisting material, and the framing braces the thin shear panels.

When code requirements for fire resistance prevent the use of combustible materi-

Figure 4.10   Formation of stud wall system with elements of formed sheet steel.

als, the stud wall can be created with a light-gage steel frame (Figure 4.10). This system is also widely used for nonstructural curtain walls on the building exterior.

4.3.2 Monolithic Walls

Walls can also be created without frames, such as walls of concrete or masonry. These will tend to be permanent walls, such as those used around stairs, elevators, vertical duct shafts, and restrooms. If these are used for vertical load bearing and/or lateral shear resistance, they are most likely to be of reinforced concrete or CMU (concrete block) construction.

Concrete and masonry walls offer several potential advantages, including the following:

1. Higher fire resistance (code rating) and general better performance in fires (no hidden fires in hollow spaces)
2. More solid-feeling; more resistive to the transmission of sound
3. Stronger walls for mounting of doors; muffling operating noises of the doors and reducing the transmitted shocks and vibrations from slamming of doors

These and other reasons may be compelling for use of the heavier construction, although choices have often to do with overall selection of the building construction systems. If a masonry or concrete structure in general is being used, these walls are the more likely to be appropriate. Exposure of the construction may also be a design feature.

To offset the previously described advantages, however, there are some problems with the solid wall forms, including the following possible ones:

1. Lack of the convenient hollow space inside the walls for wiring, piping, and so on.
2. Difficulty of future modifications of the construction (add a door, etc.).
3. Added weight of the construction; a reinforced CMU wall may weigh as much as ten times that of a stud wall, a concrete wall even more. With a lot of interior walls, this can add up fast and be a problem if seismic effects or poor soil are in the picture.

Solid walls can also be created with bricks, stone, adobe, rammed earth, gypsum plaster blocks, hollow clay tile blocks, or glass blocks. The options are especially great when structural functions are not required.

4.3.3 Panel Wall Systems

For nonstructural interior walls, a consideration sometimes critical is the ease with which they may be reconstructed or rearranged to accommodate future changes in occupant needs. This is particularly important for commercial occupancies, such as office buildings and stores.

Framed walls can be reasonably easily deconstructed, possibly even salvaging some of the framing materials. Details for the original construction of nonstructural framed partitions may be developed with this potential concern in mind.

Another option for the demountable wall, however, is the use of modular panel systems, available as priority systems from various manufacturers (Figure 4.11.) These may be full-height walls, coordinated with a modular ceiling system, and possibly even with floor systems and building services for power and communication wiring. They may also be used for partial-height, open-space planning elements in so-called office landscaping.

Figure 4.11   Interior demountable partitioning system developed for occupancies with requirements for frequent change in interior layouts.

REFERENCE

Sweet's Catalog Files: Products for General Building and Renovation, New York: McGraw-Hill, annually published, sec. 10615 (demountable partitions)

In general, choice of wall structures must relate to various overall building systems planning considerations and design criteria. Permanent walls must be coordinated in the initial design and construction of a building. Nonpermanent, demountable walls may actually be dealt with more in the category of building furnishings, especially in commercial occupancies.

4.4 WALL FINISHES

Surfacing and finishing of interior walls must respond to two general concerns: the development of the basic wall construction and the specific needs of the space for which an individual surface serves as a boundary. Finishes may be achieved naturally (as the generally untreated surface of the structure) or with applied materials. For applied finishes the range of possibilities is enormous.

4.4.1 Natural Surfaces

Natural finish may occur as the untreated surface of monolithic construction (concrete, masonry, etc.) or the generally untreated face of the surfacing materials on a framed wall. The truly raw, structural surface may indeed be acceptable or even desired in some situations. However, exposed structures often receive some special treatment beyond that provided by ordinary construction.

Structures that are intended to be covered by finishing materials are often produced with relatively rough, marginally sloppy surfaces. For a deliberate brutal appearance (dockside, warehouse, sewers of Paris look) this may be desired, and even enhanced in some cases by reworking surfaces that are too neat! Nicely finished glued laminated timbers, for example, are worked with chain saws to make them look like rough timbers.

The "natural" structure can be trimmed up somewhat, however, by simple cleaning or by some control over the basic construction process. Mortar joints in masonry can be specified as tooled; details of concrete forming can be controlled, and many factors can be given special attention. This is a situation where the designer must become very knowledgeable about specific construction practices and what can be done within reason to discipline them.

Brick construction was used in the past for many very utilitarian purposes, executed in reasonably rough fashion and covered up with plaster or other surfacing materials. Now, brick work carries a premium labor cost and is used mostly only when the true exposed appearance is desired. It is, in fact, now most often imitated with applied materials (thin brick tiles attached with adhesives, etc.), rather than created in the classic manner with mortar.

Concrete walls represent a highly manipulatable construction, which can be varied almost endlessly with respect to surface finish and detail. Basic requirements of the construction process must be acknowledged, but appearance can be altered over a great range without major change in the essential material.

Some natural finishes that may need protection on the building exterior may be left exposed or untreated on the interior, where weathering is not a problem. On the other hand, some different issues become of concern for interiors; for example:

1. Rapid flame spread or simple combustibility of surface materials
2. Resistance to abrasion, puncture, or other effects of wear or damage from contact for surfaces within reach of occupants
3. Reflective or absorptive properties of surfaces for light and sound

The appearance of interior surfaces is also conditioned by the relatively more intimate condition, as compared to exterior surfaces. Raw structural surfaces may appear better with distance to soften the texture than they will upon close examination. Actual tactile response may also be an issue with surfaces that are within reach of occupants.

For various reasons, there are many materials that are only feasible for use on either the exterior or interior, and some that can be used in either location. Plaster, for example, can be used in either location, although somewhat different forms are used for the two locations. Gypsum drywall, on the other hand, is primarily limited to interior applications, if exposed and only painted. These are applied surfacings, however, and most "natural" surfaces—such as wood siding, masonry, and concrete—can do service in either location.

4.4.2 Frame Surfacing

Wall frames may be directly surfaced with finish materials in some cases. Prefinished panels may be directly attached, for example. However, this is not the general, or most common, situation. In most cases a basic frame covering is attached and some surface finish is applied to the structural surfacing, even if only a thick coat of paint.

Frame coverings must respond to most of the concerns discussed for wall surfaces in the preceding section, including those for flame spread, combustibility, resistance to wear, and response to light and sound. The basic frame cover must also respond to structural requirements, including the spanning of the distance between frame members and any lateral shear or other functional demands.

Not a small problem is that of the means of achieving attachment of the covering to the frame. This involves a great deal of fastening and must first of all be economical. It must also respond to specific structural requirements, especially for any shear wall action. Finally, it must be permanent, so the cover stays in place and fasteners do not pop up, rust through, or otherwise intrude on the surface.

The structural covering of a frame must also respond to the specific finish materials that may be applied. This may require considerable surface treatment, as in the case of gypsum drywall, where joints must be carefully filled, reinforced, sealed, and smoothed over to obtain a seemingly joint-free wall surface.

4.4.3 Developed Finishes

Applied finishes for interior walls include a vast array of possibilities. Most of the options for exterior finishes may be considered, and indeed have probably been used somewhere. However, with the lack of concern for weather, the barriers are down and just about anything goes. Paper, fabrics, and carpeting can be used; the last often for its sound absorbency.

A consideration to be made is the concern for tactile response and general effects from contact with occupants. This may have to do with the location in the height of the wall or the character of occupants (small children, for example). Fragile materials may be used on the upper portions of tall walls, but lower parts of walls should generally have reasonably durable, easily cleanable surfaces.

Refinishing of wall surfaces without the need to disturb the basic wall structure is often a necessity, especially in commercial rental occupancies (stores and offices). Finishes that can be easily removed or covered up are preferred for these situations. A lot of carefully produced natural finishes get covered over when occupants get tired of them or don't feel like cleaning them.

Creation of applied finishes may be simple, as in the case of a simple coat of paint, or may require some additional construction, as when furring strips are used to permit attachment of paneling to a concrete or masonry wall. Interior tile, brick, or stone surfaces can be achieved with the real materials, requiring considerable preparation and adding significant additional thickness; but the more frequently used solution is to use very thin elements (real or otherwise) attached by adhesives—a much faster, thinner, and generally much more economical solution.

Wood paneling, if done with real wood at all, is usually also in the form of a thin layer of the finish wood. This may be done by using actual solid strips of the wood, attached with adhesives; by using a face ply of the wood on a structural plywood core; or by using a face ply on a thinner plywood or compressed wood fiber panel which is then attached by adhesives to some structural wall surface.

As with floor coverings and ceiling surfaces, wall finishes are often treated essentially as interior cosmetics, which can be relatively easily changed as fashions change, real user needs change, or time and wear simply exact their toll.

4.5 CEILING STRUCTURE

Ceilings are the upper bounding surfaces of interior, enclosed spaces. Structures for ceilings may be developed in a number of ways, the simplest being that of no structure at all; that is, the ceiling is simply the raw underside of the overhead roof or floor structure. However, when the exposed structure is not desired, some form of ceiling construction must be developed. The following discussions treat the common means of achieving ceiling construction.

4.5.1 Directly Attached Ceilings

Ceilings may be developed by direct attachment of surfacing materials (drywall, etc.) to the overhead structure(see Figure 4.12a). This is possible when the form, spacing, and materials of the structure provide adequately for such attachment. In residential construction, drywall is ordinarily attached directly to the underside of the closely spaced, light wood rafters or floor joists. This framing is ordinarily closely spaced to accommodate thin roof or floor decks of plywood, and the dual role of ceiling framing is easily accomplished.

When framing is more widely spaced—as when heavy timbers or steel W sections are used—it is usually necessary to use some intermediate framing attached to the major structure and spaced in response to the needs of the ceiling surfacing materials (see Figure 4.12b). Intermediate framing (furring strips, etc.) may also be necessary if the form or materials of the major structure do not permit easy attachment of the ceiling materials.

Direct attachment will yield the minimum total thickness of the ceiling plus the overhead structure, which is generally more desirable with floor construction in multistory buildings, as total floor-to-floor height will be the least. However, it will also provide the least interstitial space (between the ceiling below and the floor surface above) and may present problems where considerable space is required for building service elements, such as ducts and recessed lighting fixtures.

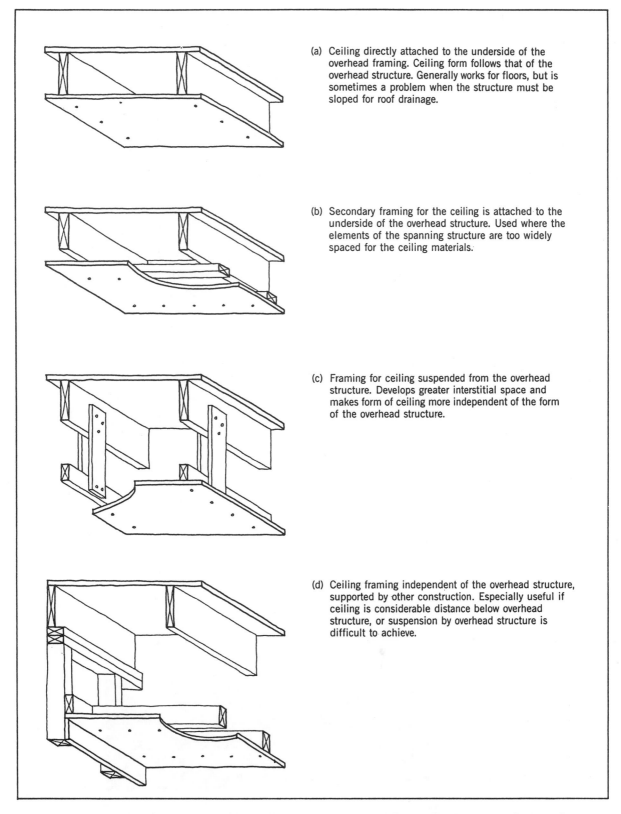

(a) Ceiling directly attached to the underside of the overhead framing. Ceiling form follows that of the overhead structure. Generally works for floors, but is sometimes a problem when the structure must be sloped for roof drainage.

(b) Secondary framing for the ceiling is attached to the underside of the overhead structure. Used where the elements of the spanning structure are too widely spaced for the ceiling materials.

(c) Framing for ceiling suspended from the overhead structure. Develops greater interstitial space and makes form of ceiling more independent of the form of the overhead structure.

(d) Ceiling framing independent of the overhead structure, supported by other construction. Especially useful if ceiling is considerable distance below overhead structure, or suspension by overhead structure is difficult to achieve.

Figure 4.12   Development of ceiling support systems: (a) By direct attachment of surfacing to an overhead structure. (b) By separate framing attached to the overhead structure. (c) By framing suspended from overhead construction. (d) By a separate framing system, free of the overhead structure.

Direct attachment of ceilings will also provide a direct reflection of the general form of the overhead structure—not so much a problem with floors, but possibly one with nonflat roof structures. For this or other reasons, it may be desired to produce an essentially independent ceiling structure, as discussed in the following sections.

4.5.2 Suspended Ceilings

A separate ceiling structure may be hung from the overhead structure, deriving support from it but not necessarily reflecting its profile or detail (see Figure 4.12c). This is often done to create needed space for equipment and services, but also to create a different form or simply a lower ceiling level.

Suspension must be achieved with materials and details that relate to both the overhead structure and the form of the ceiling construction. Typical methods and materials used relate to common associations of structures and ceiling constructions.

Many ceiling surfacing materials can be used in this situation, as well as with direct attachment to the overhead structure. Intermediate framing may be somewhat different if it is suspended instead of being directly attached. Suspension elements can be made independent of the spacing of the major structure, such as when hangers are attached to the decking instead of the framing—a common practice with decks of steel or concrete.

Many priority modular ceiling systems are developed primarily for use in suspended construction (Figure 4.13). These ordinarily have modular surfacing panels and a network of framing, and may even include special hangers. They are often also coordinated for use with modular lighting and HVAC units in a fully integrated system.

Because they exist often to provide for service systems overhead, suspended ceilings may need to provide for access to the elements of the enclosed services. This sometimes favors the use of the priority, modular unit systems, whose individual surface units are usually quite small and readily removable.

Figure 4.13   Modular ceiling system with suspended framing and lay-in units. Elements for lighting, HVAC service, fire sprinklers, smoke detectors, and so on may be developed within the modules, but often also fall randomly within the system.

4.5.3 Separately Supported Ceilings

In some situations, ceilings may be developed with construction that is totally independent of that overhead (see Figure 4.14d). This may simply be the practical means for achieving the ceiling, or it may be necessary where the ceiling is a great distance below the structure above it.

Where room sizes are small, and wall construction permits it, a ceiling may be developed with separate framing supported by the room walls. This framing may be quite modest, if it only needs to support the ceiling and does not provide for a floor for the space above it. Design live loads for ceiling spaces with limited access are usually only 10 psf. Span limits may be derived from critical concerns for deflection (visible sag) rather than from actual load-carrying capacity.

For some types of occupancies—notably speculative, commercial ones, such as offices and stores—use of independent interior wall and ceiling construction allows an increased level of freedom and ease for modification of interior spaces.

4.6 CEILING FINISH

Ceilings are considerably in view, so they deserve considerable attention where the quality of appearance of the building interior is of concern. This attention must be given to the ceiling construction in general, but mostly to the finishing of exposed surfaces.

4.6.1 Exposed Construction

The simplest ceiling finish is no finish at all, which may occur if the overhead structure is left exposed. However, as discussed for natural finishes on walls in Section 4.4.1, the unadorned and untreated ordinary structure may not be very appealing. Cleaning, a coat of paint, or other minor dressing up may be in order. More seriously, there may be some need to specify some greater care in the construction, in the selection of materials, in careful arrangements of parts that are not ordinarily treated with much precision of alignment, and so on.

The decision to expose the structure may be a practical one, as is done for garages, industrial buildings, chicken coops, and other not so glamorous interiors. It may also be a very conscious design decision for a relatively high-class interior (Figure 4.14) where a very dramatic structure is deliberately used and exposed in a high-tech style. In the latter case, there may be a desire to be very honest to the ordinary details of the proper construction, or some considerable design effort may be expended in developing a much more elaborate or generally smart-looking structure.

Some structures may not endure exposure well and need some real protection, especially for fire resistance. The ability to leave structures exposed—especially wood and steel ones—may be qualified by various conditions pertaining to the building occupancy, size of the building, height of the structure above the floor, and so on.

Reasonably raw exposed structures on interiors are parts of some very dramatic and exciting historic architecture, such as the Gothic cathedrals; the great train sheds of the early industrial period; Paxton's Crystal Palace; the richly formed concrete roof structures of Nervi, Candella, and Torroja; and the convoluted tension structures of Frei Otto. One cannot imagine having a suspended ceiling beneath such awe-inspiring displays of structure.

Figure 4.14   Ceiling developed essentially as the exposed overhead construction; cleaned up, but generally untreated for ceiling purposes.

4.6.2 Drywall

The ubiquitous drywall panel accounts for the vast majority of ceiling surfaces in present construction. This is, in general, the all-purpose cover-up, forming both vast, uncluttered flat ceilings, and boxing around various complex, coffered, coved, curved, and various multifaceted forms.

There are indeed many types of drywall panels and many different special details for their installation. Modifications may be made for increased fire resistance, greater strength, better sound response, or other desired properties. Special panels may be used for exterior sheathing, plaster bases, reflective insulation, and other purposes.

Fastening of panels to supporting framing and treatment of joints are major detail concerns, also having many variations. Long experience has resulted in some very standard forms of construction, but the technology continues to evolve.

As with any material or form of construction, drywall has its limits, leaving some situations where other construction is clearly indicated. However, after its long, hard-fought battles with trade unions, code writers, and insurers, and its emergence from the stigma of being a cheap imitation of real plastered (wet-wall) construction, drywall is now pretty much the definitive wall and ceiling surfacing material, to be supplanted only for cause in special situations.

4.6.3 Plaster

Plaster is a very ancient material, originally developed primarily as a coating and protective covering for soft, vulnerable structures, such as adobe and soft-mortared brick or stone masonry. In many cultures, the smooth plaster surface became a canvas for painters, producing some very rich interior adornments. Plaster also became a sculptor's medium, producing rich forms, such as those in the baroque period.

As with many historic materials, plastered construction is now often imitated, rather than real. Even stucco (exterior plaster, although still extensively used, is now also frequently imitated in plastic-based finishes on multilayered exterior insulation systems with fiberglass reinforcement and foam plastic backup.

However, real plaster can still be installed, pretty much in the traditional, "wet" method. A very high-quality, durable finish can still be produced with a multicoat plaster finish, ending with a very fine-textured, hard surface. It may be expensive and time-consuming, but will have properties that are hard to compete with.

For practical purposes, plaster is now sometimes installed over perforated gypsum wall panels, becoming transitional as an independent wall material and marginal between old-style plaster and merely a very thick coat of paint. Some form of backup is generally needed, and is now likely to be either a formed steel product or some composite of paper, moisture-resistive coating, and steel reinforcement.

Gypsum plaster—the same material used in drywall—can be installed wet, but is too soft in general to be used for surfaces within reach of occupants or otherwise subject to contact or wear. It is, however, very low in density (lightweight) and may be a good choice for ceiling surfaces.

Really hard plaster is basically portland cement-based and forms a very dense, hard material. Stucco, applied in a reasonable thickness and reinforced with steel wire, literally forms a thin reinforced concrete slab unit with major contribution to lateral stiffening of a building, particularly one with a light wood frame.

Sculpted while wet, plaster surfaces can be given various textured forms besides dead flat and smooth. Other materials can also be worked into the surface just before hardening, such as paint (fresco), small tile, grit, crushed glass, and so on. Plasterers were once prima donnas of the building crafts, but the craft is now largely nonexistent.

4.6.4 Modular Systems

Modular systems may consist of units adhered directly to the underside of a concrete slab or to an untreated gypsum drywall surface. More often, however, modular systems occur as a priority system using separately supported frameworks and small lay-in panel units (Figure 4.13).

Modular systems typically use some dimensions that relate to other ceiling items, such as light fixtures and grilles or registers for HVAC systems. Some priority systems are fully developed to incorporate these service elements in a totally integrated manner. Many commercial interiors are developed with these systems.

The dimensions and repeating modules of a ceiling system may be essentially divorced from those of the structure from which they are suspended. However, when interior partitioning must intersect the suspended ceiling, coordinated wall/ceiling systems may be used. There are many alternatives for this situation, including ones emanating from ceiling system manufacturers and wall system manufacturers. However, custom designs are not all that difficult either. Most modular ceiling systems can be trimmed to fit at their edges to accommodate some nonmodular wall construction or other special construction situations.

If a modular ceiling is to be used, however, it will come off much better if everything related to it is kept neatly in the modular order, including lighting, HVAC elements, fire sprinklers, audio systems, signs, and any decorations.

4.6.5 Miscellaneous Ceiling Finishes

Being typically out of reach, ceiling surfaces may be developed with just about any material—and probably have been. Wood, metal, plastic, paper, and fabrics are common. Egg cartons, growing plants, and glass are not unknown.

Fire codes and reasonable safety must be considered, but the highly visible ceiling surface is a blank canvas for the interior designer, and fair game. Conventional finish materials, used for interior or exterior walls and even for floors, may be used. This is probably the safest place for designers to cut loose, as the fewest functional constraints exist.

As with any interior surface, various physical properties must be considered, however. Many of these are discussed in Chapter 5.

4.7 DOORS

Exterior doors serve to control entrance to the building and generally are primary fire exits. Major features in their design are needs for security and weather resistance. Interior doors generally provide access to selected interior spaces and have specific needs depending on the spaces they serve.

Basic uses and components of doors are described in Section 2.5. In this section, some considerations for selection of doors are presented.

Most doors are selected from catalogs or directly from ordinary stock of building materials suppliers. However, many options are possible regarding hardware, trim, and special features, so that a single door often requires several considerations for a complete specification.

4.7.1 Types of Doors

The type of door selected—as well as its size, shape, materials, and general operation—must be considered for the specific usage involved. Traffic through a door may be limited to people, but also frequently involves movement of vehicles and furnishings or supplies related to interior spaces. Requirements for operation and access by handicapped persons may provide some critical design criteria.

Most small doors swing horizontally, with hinges on one vertical edge and a latch on the opposite edge to hold them shut. However, many other forms are possible.

When doors move, they must move to some place other than their closed position. This presents various problems for planning or other concerns. Swinging doors may swing into people on the opposite side, into furniture, or into a position blocking some traffic or the operation of another door. Door swings are ordinarily shown on architectural plans to illustrate the spatial needs for their operation. Swings may be restricted in some cases by bumpers or other devices.

Door dimensions ordinarily vary through a standard range of sizes, relating to available stock sizes, but basically deriving from code requirements as well as practical applications. Code-minimum sizes are often used, but larger sizes are also used to dramatize major doors or relate to the scale of the door placement.

A critical set of sizes derives from the minimum dimensions required for fire exits. This affects both height and net width of the clear doorway provided when the door is swung fully open. Net clear width may also be determined by requirements for access by persons in wheelchairs.

The side-hinged, horizontally swinging door is the most common form. This is the type normally required for fire exits, for example. Sliding, pivoting, rotating, folding, upward-swinging, rolling, and other forms of operation may be possible for some situations, but practical usage and code requirements should be carefully considered.

4.7.2 Door Components

As described in Section 2.5, doors ordinarily have basic units that include a frame, the door, and the various accessories and features that facilitate its installation and operation. For a door selected from a catalog, the full kit may include almost everything required, or it may be limited to the door itself. Complete packages are somewhat less common here than with windows, since doors have considerably greater variability, particularly with regard to hardware and accessories.

A given building may use a number of doors, with many doors essentially the same in appearance. Yet practically every door may have *something* different: direction of swing, restriction of swing, type of latch or lock, automatic closure, hold-open device, peep hole, kick-plate, glazing, and so on. Doors also have two sides, and one side may be the same for many doors, while separate room interior conditions change, in interior decoration, if nothing else.

Coordination of the doors, with all of their differences, is a major design task. Each item for each door is an individual design decision that must be an informed one.

4.7.3 Door Operation and Features

Door operation can be very elementary—just push it open, for example, as is sometimes done for kitchen doors or the old-fashioned saloon or cafe doors. It can also be quite complicated when many different concerns must be provided for simultaneously: fire safety, security, handicapped access, sound privacy, and a pleasant, welcoming effect, for example.

Complete operation frequently involves almost every component of a door. Construction of the frame must relate to the form of support required and the installation of hinges, latches, locks, and so on. The door itself must facilitate hardware, resist handling, incorporate glazing for viewing, and receive any special accessories for swing restriction, automatic operation, and so on. And, of course, much of the hardware in general relates to operation functions.

Beyond the door itself are various architectural design considerations that affect the decision to have the door in the first place, where it should be placed, what form it should have, and how it should be operated. Unlike windows, doors are seldom used strictly for design effect; they generally provide access, exits, traffic control, fire barriers, and many specific functions.

Door usage may also require some extra items besides the door and its basic hardware. These may include lighting, signs, signaling, and various elements that deal with the control or use of the door and the passageway it represents.

Finally, doors are parts of walls, and must provide many of the wall functions, such as acoustic separation, fire barrier, and so on. Making them operational while also meeting the demands for the solid wall can be quite challenging.

4.8 STAIRS

Stair construction usually relates to the general form of construction of a building. There are two basic reasons for this. The first has to do with simple compatibility of the construction—a wood stairs makes sense in a wood-framed building, for example. The second reason has to do with the general requirements of a building for fire resistance; this has its primary effect on the basic choice of construction for the building and then extends to a compatible set of requirements for the stair construction.

General design concerns for stairs are discussed in Section 2.6, and some basic planning issues are illustrated in Figure 2.7. Many building stairs are fully enclosed by walls and accessed through doors—a general requirement when they must function as fire exits. In fact, although they would seem to be logical necessities in multistory buildings, most stairs are planned for in the first place because they are required by the code as fire exits. While elevators may normally carry the bulk of intralevel movements of people, they do not constitute legal fire exits and must be supplemented by stairs.

In some situations, however, stairs may be open—that is, not enclosed by walls. They may also be permitted to be circular or winding in plan, have open treads, be suspended, and so on. Building size, occupancy, stair placement, and what spaces the stairs serve may determine these possibilities. Some very beautiful stairs have been developed for situations where these freedoms occur. Their basic three-dimensional character can make them become major sculptural elements in a building interior.

Required fire stairs are usually developed as very Spartan, unadorned, minimal construction; typically detailed in close conformity to minimum code requirements for dimensions and features. A very high-class building may have really dazzling elevators, as they mostly serve as the vehicle for building users. But close by will be a fire stairs of very modest design. The door and sign may be high style, but once inside construction is humble.

Fire stairs are often constructed with priority systems designed to be slipped into a void in the building construction. The walls and the landings at the floor levels may be parts of the general building construction, but the stairs themselves, plus intermediate landings, are a custom-designed unit, supplied by a specialty contractor.

For any purpose, stairs must be designed to conform to some safety standards that affect the angle of the stair (usually expressed as the riser/tread ratio), limits on actual riser and tread dimensions, net clear width, overhead clearance, maximum length of a single flight, and requirements for handrails, railings, lights, and so on. Special consideration should be given to any requirements for users with diminished capacity (very young, very old, infirm, handicapped, blind, etc.).

Stairs are a universal requirement, and many of the common methods for producing them are quite standard. Nevertheless, the complete planning, design, and construction detailing of stairs is a major interior design problem.

REFERENCES

Mechanical and Electrical Equipment for Buildings, 7th ed., New York: Wiley, 1986, chap. 13. For fire requirements.

Architectural Graphic Standards, 8th ed., New York: Wiley, 1988.

5

Special Concerns

The initial process of building design generally involves the determination of the shape, dimensions, and general arrangements of interior spaces. Next comes the basic determination of the means for achieving the construction in terms of materials, systems, and details. Eventually, accountability of the design must be acknowledged with regard to various special concerns. This chapter considers some of these major concerns.

5.1 SURFACE AND BARRIER ENHANCEMENT

Development of interior surfaces must respond to various concerns for appearance and maintenance of surfacing materials. Surfaces also relate to the way in which light and sound are experienced by building occupants. As discussed in Chapter 2, the development of surfaces and of the interior construction in general is, in many instances, modified by considerations for enhancement of various properties of the construction.

Basic forms of construction have particular properties with regard to various physical responses, each being more or less fire resistive, more or less resistant to sound transmission, and so on. Achieving better levels of response may be accomplished by simply choosing among common alternatives for the construction. However, it is also possible to effect various modifications to enhance the ordinary forms of construction.

In the typical situation, many different properties of the construction must be considered. The optimal solution for any single response is not likely to be optimal for all other designed responses. What is cheapest is not likely to be most durable, most attractive, most fire resistant, and so on. Primary values must be given to important design goals, and some other factors must be made less critical.

Some primary behavioral responses are considered in the following discussions.

5.1.1 Fire

The single behavioral concern that most affects choices of construction for building interiors is fire. This is the source of a great deal of what constitutes building codes; choices concern use of materials, selection of basic systems, and dimensions and details of the construction. The fundamental considerations from which code requirements are derived include the following:

1. Combustibility: Just about anything can be consumed or otherwise destroyed by fire. However, some materials and elements, such as thin pieces of wood or

paper, can be easily ignited and quickly burned. To the extent that parts of the building construction can add to the so-called *fire load* (total mass of combustible materials present), the construction represents a potential hazard on its own—regardless of the building contents or activity of occupants. Reducing this hazard constitutes a major design contribution to fire safety.

2. Vulnerability: To the extent that the construction is vulnerable or susceptible to fire, and likely to be easily or quickly made nonfunctional (for example, collapsed), it is considered to lack resistance to fire. This is especially of concern for parts of the construction that provide structural support or effect separation for portions of the building. Floors and walls should neither quickly collapse nor quickly permit fire to spread by penetrating them.

3. Toxicity: Most combustible materials release some toxic (deadly, life-threatening) materials when they burn. They also consume oxygen to produce a fire. The greatest loss of life in building fires is due to either oxygen depletion or inhaling of toxic materials (smoke, poisonous gases). Choice of materials is critical in this regard.

4. Construction form and details: Various aspects of the construction and building planning may affect fires or the safety of occupants and fire fighters. Suspended ceilings are among a fire fighter's worst enemies; fires can occur within their enclosed spaces and spread rapidly without being detected, bursting out suddenly or causing the collapse of the ceiling or the structure above without warning. For doors, their location, visibility, size, direction of swing, and ease of operation relate to the potential use of the door as a safe fire exit. Dead-end corridors (no way out, except the way you came in) are firetraps; the longer the corridor, the more deadly.

Most building code regulations regarding fires were initially developed and promulgated under the sponsorship of insurance companies. Insurers typically cover the building, its contents, and its occupants—all being of concern for settlements of claims. Most building fires are relatively quickly contained and extinguished without loss of life or serious injury to occupants. Still, even small fires cause damage to the building and its contents, and cost the insurers money. Insurers are not to be viewed as bad guys, but they are basically in business to make money, not to function as humanitarian organizations.

Making life safety the prime goal, rather than simply reduction of insurers' losses, is just beginning to effect changes in building codes. Fire sprinklers, for example, cause a lot of damage on their own, but are very effective for life safety. Real effectiveness for life safety and for quick, practical, and safe fire fighting is ascending over the concept of protecting the building itself and/or its inanimate contents.

In this as in other regards, it should always be borne in mind that building code requirements are typically minimal, not optimal. Optimization of conditions for life safety essentially implies some levels of protection and behavior enhancement over the code minimums. Try asking a building design client if he or she wants a minimally safe building.

5.1.2 Sound Control

One of the most elusive design factors is that of sound control. Goals for this differ for different occupants and for individual interior spaces. Typical major concerns are the following:

1. Privacy: This generally means the keeping of sound—notably conversations—within a contained situation, frequently simply inside a room. Normal conversation has a specific range of sound level (decibels) and frequency (cycles per second), so that either transmission through boundaries (walls, floors) or masking with other sounds (music, etc.) can be reasonably managed in design.
2. Noise intervention: Noise is essentially unwanted sound—it could be beautiful music, but intrusive if it drowns out a lecture or conversation or keeps someone awake who is trying to sleep.
3. Room acoustics: This has to do with the management of sound within a bounded space, involving sustained, ambient sound levels (enduring sound), reverberation, echoes, and amplifications of specific frequencies.

Controlling any or all of these, or other special problems, may be reasonably established by careful design. Effecting it in the actual construction is another matter. Like air or water, sound can leak through tiny openings. It also travels by paths you don't anticipate—like out one window and in the next, instead of through the strong barrier you have created between rooms.

Effective sound management requires a lot of cooperation between architect, acoustical engineer, builder, and even building occupants. That is a lot to ask for, but any one of these parties can subvert the effort.

Many simple things can be done in choosing surfacing materials, detailing of the construction, shaping of interior spaces in planning, isolating of noisy equipment, and other situations to generally improve sound conditions. Classically dumb situations can be carefully avoided by minor attention in dealing with design.

It is possible to throw a lot of money at sound control and miss the target. Major efforts must be made for serious situations, like those of auditoriums and recording studios. For most buildings, however, some simple attention to potential problems and the easy things to do to avoid them or compensate for them are the usual course.

One simple trip through a basic reference on good sound management in buildings will sensitize the average designer sufficiently to the simple issues. A lifetime of experience will then produce some judgment about what is really possible to achieve in ordinary circumstances.

REFERENCE

Mechanical and Electrical Equipment for Buildings, 7th ed., Stein, Reynolds, and McGuinness, New York: Wiley, 1986, chaps. 26 and 27.

5.1.3 Security

The exterior, enclosing construction of most buildings must offer some degree of security, referring mostly to the prevention of intrusion by unwanted persons. This varies considerably, of course, from the situation of a residence in a peaceful community to a bank in an urban area. Security involves the basic construction of exterior walls and roofs, but also the nature of doors, windows, skylights, utility tunnels, and all other potential points of entry.

Individual interior spaces or levels in buildings may also need to be made secure; the basic need and degree of security required depends on the building usage. Determined burglars are hard to thwart. Most security achieved by the construction alone is in the order of preventing casual entry. Serious security—in construction terms—is usually reserved for jails, bank vaults, and other special situations.

As with design for other building responses, some simple considerations can improve security without major restriction on the general construction or general building planning. Good lighting at entries, windows and door locks that are resistive to forced entry, and other measures can improve the situation. In the end, however, alarms and surveillance are probably the real effective means to establish security, and effective, economically feasible measures within the scope of building design of limited value.

REFERENCE

Mechanical and Electrical Equipment for Buildings, 7th ed., Stein, Reynolds, and McGuinness, New York: Wiley, 1986, chap. 22.

5.2 BARRIER-FREE ENVIRONMENTS

Increasing attention has been given in recent times to the development of buildings that are safe and are accessible by persons of limited capabilities. Design of public facilities has been most affected, but access to workplaces has also been promoted, so that many basic elements of construction are now affected by requirements in this category.

Early on in this effort a principal concern was for access for persons confined to wheelchairs. The basic concept and general interest is now of a broader scope and tends to spill over into general concern for a much wider range of building users, beginning with those with serious limitations, such as blindness, but extending to the elderly, the very young, illiterate persons or others incapable of reading signs, and so on.

Specific details and features of buildings can be made less hostile or otherwise more accommodating to users—those with limited capabilities or not. Elimination of steps where possible, widening of doors, provision of grab bars in toilets and showers, and use of large, clear letters and identifying symbols on signs represent efforts to create a generally better environment for everyone; but surely they do indeed widen the range of potential building users.

5.3 LIFE SAFETY

Attention to life safety has a simple goal: reduction of the potential for loss of life or for serious injury to building users. This is a major factor in the general development of building codes, which are government ordinances when enforced—providing essentially for the public health and welfare.

Building codes are, however, basically meant to establish threshold, minimal levels of acceptability. *Optimizing* for life safety, without necessarily creating economically unfeasible or otherwise distorted designs, is a relatively new design emphasis that is gaining strength. Viewing all aspects of design decision making in the building design process from this point of view, and having evaluative data to input to a value analysis procedure, is a major research field in building design.

5.4 SIGNAGE AND TRAFFIC CONTROL

Buildings are essentially inanimate, although the operation of various building services—elevators, HVAC systems, water heaters, and so on—involves some dynamic concerns. People, on the other hand, come and go, move about inside the building, and exit—usually reasonably leisurely, but occasionally in a rush, when there is a fire alarm or bomb scare.

Longtime users of a building know the premises and how to get around in the building. Infrequent users or first-time visitors often need help, starting with finding the entrance.

Design of the building construction is not the way to deal with traffic management, but various considerations for the building usage may well affect the general planning of the building and the choice of some elements of the construction. Doors, for example, are major elements in traffic control, and many of their features derive from these concerns.

Some designers measure really effective planning by the lack of need for signs. This may be a good ideal, but some signage is actually required by codes, and good signage is never a negative thing for building users.

Visibility within the building interior may be a means of assisting users to find places. Walls of glass or doors with glazing may be effective for various purposes. More subtly, continuity of wall construction or simply wall or floor finishes may keep people on a traffic path by inference. A sudden change—in the size or shape of bound spaces, in lighting, in visibility of surrounding spaces, in floor finish, and so on—can grab peoples' attention and alert them to something.

Effective exiting of the building in emergencies should be carefully planned. This may begin with minimum code requirements, but should extend to deeper concerns for practical likelihoods of the responses of users. Fire exits may be legally marked by signs, but if nobody enters the building through them, they are not going to rush to them in an emergency. Safety and general security of interior construction should preferably be progressively more stringent as exiting occupants proceed along logical exit paths.

6

Systems

System is a much overused word, but generally it describes any group of component elements that exist in some ordered relationship. Most often the group is composed to relate to some specific task. A building as a whole entity can be described as a system, although it is more common to use the term to describe some subunit, or subsystem, of a building. In this chapter the issue of building systems is discussed with reference to the idea of a particular, distinctly definable means of construction. In that context, there are a small number of very ordinary systems that are repetitively used to create buildings.

6.1 HIERARCHIES

Systems with many parts often have the parts arranged in some ranked order or hierarchy (see Figure 6.1). For building construction a hierarchy may be defined in terms of the relative importance of elements in reference to some specific task or issue. Individual parts may be viewed as major or minor, primary or secondary, and so on. This ranking may relate to considerations of the principal determinants of the system or to the relative significance of parts to design modifications for a specific purpose.

Modifications in the form of design variations may be more easily done when they deal only with secondary or minor elements of the basic system. Thus, with the light wood frame, the studs are primary determinants of the system, while surfacing is secondary to the basic system. Many types of materials may be used for surfacing—effecting major changes in the appearance and various properties of the construction—without essentially altering the basic nature of the structure.

The quality of adaptability just described for the light wood frame is one of the reasons it sustains its popularity; it can indeed adjust to almost endless variety in terms of application to different design styles and situations. There are a very few enduringly popular, basic construction systems that enjoy this quality.

In considering the major basic systems described in Section 6.6, the reader should visualize the hierarchies in them and the degree to which significant architectural design modifications can be made without altering the basic character of the systems.

6.2 DESIGN PROCESS AND TASKS

A major early design task is the selection of the basic systems of construction for the building in general and for its major elements: roof, walls, floors, and so on. Options may be innumerable or may be shortened considerably by various design criteria,

Figure 6.1   Hierarchical system for definition of a building structure. Rolled steel W
sections create basic structural frame of columns and column-line beams.
Nonstructural curtain walls and partitions infill vertical wall planes, with a
wide choice of construction, due to their nonstructural nature. Floor joist
and deck system infills the horizontal planes of roofs and floors, also with a
wide choice of construction—in this case, composite lumber and plywood
I-beams, plywood deck, and concrete topping on floors.

such as the construction budget, time available for design and/or construction,
building code requirements, or strong client preferences. However determined, the
firm choice of basic systems at an early design stage will greatly simplify the progress
of the design work.

With basic systems defined and their hierarchies understood, the forms of design
modification that are most easily achieved can be explored. If the major design goals
cannot be achieved within practical limits of the basic systems chosen, it is best to
find this out very early. A shift in the basic systems has major effects on the progress
of the work, and the earlier it happens, the better.

Although real design work seldom proceeds in so orderly a fashion, the following is
a logical process for determination of the construction during the general development of a building design project.

1. Determine the basic systems. This means the selection of general forms and
   materials (masonry bearing walls with CMUs, for example).
2. Determine specific features of the basic elements (mortar type, basic unit type

and dimensions, typical details, and other critical specifications for the CMU walls).
3. Consider effects on major building elements. This anticipates various problems of design integration. (How do unit dimensions affect window and door sizes, room dimensions, etc.?)
4. Develop all necessary major architectural details for use of the system. This tests the variability of the system and the ability to generate the range of situations that must be accommodated in the building, as well as the potential for alternatives.

6.3 MIXTURES

Buildings consist of large collections of individual parts and subsystems. There are very few manufacturers of whole buildings; in the main, each manufacturer produces only selected parts. The task of determining the collection of parts for a particular building is up to the building designer.

Practicality generally dictates that the parts of buildings are mostly obtained as manufactured products, selected from the catalogs or samples of the products as supplied by manufacturers or suppliers. In this situation, the building designer has the major task of comparative shopping and mixture making.

While a building represents a giant mixture, it is usually a mixture of subsystems and components, each of which may also be a mixture. For example, a designer may choose to use a light wood frame (2 × 4 wood studs, etc.) for a wall. This is a well-defined basic construction component, and it can be "mixed" with various other components for an appropriate building assemblage. But the stud wall itself is a mixture, with studs produced by one maker, surfacing by another, connectors by a third, and so on. Each item can represent a design choice not necessarily implied by the general choice to use a wood stud wall system.

Selection between the available products for a single use (surfacing for a stud wall, for example) requires comparative value analyses that individual suppliers cannot be fully relied upon to assist with an unbiased view. The larger view of options, as well as real freedom of preference, is required. One type of wall surfacing may be the best in its class, with many superior qualities, but those qualities may be of minor concern for a particular application, and a lower-quality, less expensive product may be perfectly adequate for the limited task at hand.

Designers (shoppers, buyers) must look diligently for the best choice for each individual part for a building. But the whole building is more important than any individual part. The best roofing material, placed over a poor roof structure, will not produce a good roof.

6.4 APPROPRIATENESS OF SYSTEMS

The achievement of good construction begins with the choice of the appropriate materials, products, and systems for a particular use. And in careful detailing, proper and thorough specifications, and competent construction work, and a well-built building just might result. But a well-built building in the wrong place or for the wrong use will still be inappropriate.

It is one thing to put plaster on a wall in the right way; it is another to select plaster as the appropriate surfacing for the wall for the situation at hand. Product information and standard specifications can deal with the correctness of the plaster itself, but the larger question of whether or not to use plaster is in the hands of the designer.

Manufacturers or industry associations sometimes provide guidance to the appropriate use of their products. However, they have in general a major interest in promoting the use of their products, so they do not much like pointing out all the situations in which those products are inappropriate.

Building codes provide some guidance for the selection of materials and systems for appropriate situations. This occurs mostly as a negative form of criteria—prohibiting certain uses in particular situations. These should be determined for any design work and used to narrow the choices very early on, but they typically leave room for many options.

Various unbiased reference sources exist for some assistance in making appropriate, unbiased choices. However, individual buildings always have some specific qualified conditions that make simple evaluations difficult. Timeliness is also a major concern, as well as the many influences of regional situations.

In the end, the observations, experience, and judgment of the designer must prevail. This is what design is all about, and the task is humbling.

6.5 CHOICES: THE SELECTION PROCESS

Design of building construction is usually simpler and faster to achieve when choices can be made in large bites. Choosing preestablished (or essentially predesigned) systems is one means for doing this. If the choice is good in all regards, the size of the bite is irrelevant.

Preestablished systems offer the potential advantage of some demonstrated success. If the demonstrated successful cases have a reasonable match with the situation for a proposed design, all the parties concerned—designer, builder, owner, investors, insurers, and code-enforcing agencies—may take some comfort in an assurance of success. This is, of course, the actual case, making any real innovative design difficult to sell in many quarters. Designers should not be totally discouraged, but must understand the realities in achieving any significant changes in the "good old ways" of doing things.

Exciting, uplifting architecture can really be achieved with very ordinary means of construction, but can also—on occasion—come from real innovation or dramatic usage of the construction. In either case, the designer usually has a considerable understanding of, and appreciation for, the construction details and processes. If ordinary means are used, the designer understands their limits and potential, as well as what significant architectural design variations can be achieved within the basic systems. If ordinary means are rejected, the designer usually does so in full knowledge of what exactly is being discarded and what the needs are for any replacements. Otherwise, innovations will inevitably be naive and quite likely unsatisfactory in some performance aspects.

6.6 GENERAL SYSTEM DEVELOPMENT

Current construction of buildings in the United States occurs with a broad array of materials and systems. Some are modern versions of very old methods and uses; some are very recent additions to the inventory of building technology.

The majority of buildings—modest in size, but extensive in number—are built with a few common, basic systems. Many others are built with only minor variations of the basic systems. The larger the building, the higher the total construction budget; or the more prominent its place in the community, region, country, or possibly the client's esteem, the more likely its design budget may support some real innovative design effort. Otherwise, innovation becomes mostly a determined personal effort of individual designers, probably mostly on their own time.

The following discussions present some of the most common methods for achieving interior systems (mostly floors and interior walls) for buildings. Specific examples of most of these are also shown in the building case studies in Chapter 7.

6.7 FLOOR SYSTEMS

Floor structures can be executed in wood, steel, or concrete, and various combinations thereof. Many factors will influence the choice of the basic structure, including code requirements for fire, design live loads, required clear spans, general occupant needs, and the general construction system for the building.

A common floor is the concrete paving slab, as discussed in Section 4.1.1. Other floors consist of systems that begin with some spanning structure, adding the necessary finish on top and developing whatever may occur beneath the structure. The following sections present the most common solutions for this type of system.

The underside of a spanning floor typically constitutes the ceiling for a space below. The full development for a floor in this case includes the structural development of the ceiling and usually some accommodation for building service elements in the interstitial (within the structure) space between the floor above and the ceiling below. Some of these concerns are addressed in the following discussions, but complete systems are more fully illustrated in the examples in Chapter 7.

6.7.1 Light Wood—Joists Plus Plywood Deck

The floor structure using closely spaced, solid sawn wood joists with a plywood deck is firmly entrenched as the most common building system for modest-size buildings in the United States. Some typical details for this system and some of its typical characteristics are described in Figure 6.2.

The basic system, as shown in Figure 6.2, is used extensively. However, it is also possible to modify the system in various ways, using alternative elements for the joists and/or the deck. Some of these alternatives are discussed in following sections.

The simplest ceiling structure, most commonly used, is the gypsum drywall panel system, with panels nailed directly to the bottoms of the wood joists. This allows for some interstitial space use between joists, but none across the joists; thus it is also possible to suspend a ceiling, if necessary, typically using hangers from the joists.

The plywood deck is rather light and the overall floor does not have a very solid feel. It is also not so great for resisting sound transmission. For these and possibly other reasons, it is increasingly common to use a thin layer of concrete fill on top of the plywood deck, especially when different parties occupy the spaces above and below the floor structure (motels, offices, and apartments, for example).

The plywood used is quite thin, and all of the edges of the panels typically need support. This may be provided in various ways, the most common being the use of solid blocking (short pieces of joists) between the spanning joists or by using thicker plywood (7/8 in. or more) with tongue-and-groove joints at the edges.

Use of this system is illustrated in the building examples in Sections 7.1 and 7.2. The principal reasons for *not* using this popular system are its lack of fire resistance

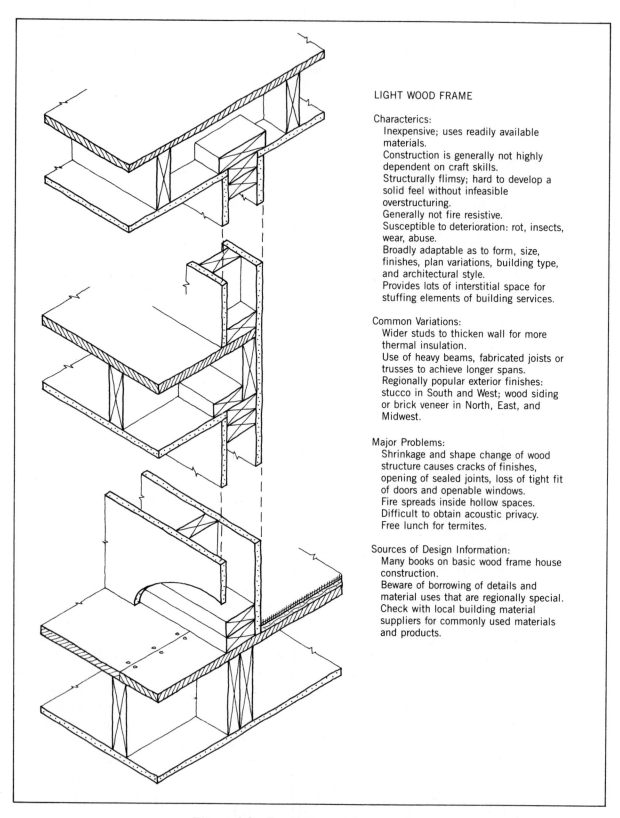

LIGHT WOOD FRAME

Characerics:
  Inexpensive; uses readily available materials.
  Construction is generally not highly dependent on craft skills.
  Structurally flimsy; hard to develop a solid feel without infeasible overstructuring.
  Generally not fire resistive.
  Susceptible to deterioration: rot, insects, wear, abuse.
  Broadly adaptable as to form, size, finishes, plan variations, building type, and architectural style.
  Provides lots of interstitial space for stuffing elements of building services.

Common Variations:
  Wider studs to thicken wall for more thermal insulation.
  Use of heavy beams, fabricated joists or trusses to achieve longer spans.
  Regionally popular exterior finishes: stucco in South and West; wood siding or brick veneer in North, East, and Midwest.

Major Problems:
  Shrinkage and shape change of wood structure causes cracks of finishes, opening of sealed joints, loss of tight fit of doors and openable windows.
  Fire spreads inside hollow spaces.
  Difficult to obtain acoustic privacy.
  Free lunch for termites.

Sources of Design Information:
  Many books on basic wood frame house construction.
  Beware of borrowing of details and material uses that are regionally special.
  Check with local building material suppliers for commonly used materials and products.

Figure 6.2   The light wood frame system.

and its limited span capabilities (typically 20 ft or less). The latter may be overcome by substituting other elements for the solid sawn joists (typically a maximum of 2 × 12, in most cases), but this begins to produce a different basic system, some of which are described in sections that follow.

This system is mostly used with an all-wood structure, but mixture with other systems is also possible, as discussed in Sections 6.9 and 6.11.

6.7.2 Heavy Timber

A second form of all-wood floor structure is that described as *heavy timber.* This specific term is used to qualify a system that is given a specific fire rating by building codes, due to the relatively slow-burning character of the structure. For code qualification, supporting beams must be at least 5 in. thick and decks at least a nominal 2 in. thick. The basic form of the system and some of its characteristics are described in Figure 6.3.

The larger members used are typically spaced considerably wider than the joists in the light wood system (Section 6.7.1). Spacing limits relate to the span capability of the deck. The minimum deck, typically slightly less than 1.5 in. in actual thickness, can usually span only about 5 ft with typical loads. Thicker decks are possible, and sometimes used, but the common use is with the thinner plank units—of either solid sawn wood or laminated form.

Timber members may be solid sawn or glue laminated, the latter being generally required if spans are great. This is a practical, utilitarian system for roofs, especially where the structure is exposed to view and the system is architecturally appropriate to the interior design. It is less used for floors and is more often "faked" with decorative members developed as a ceiling finish.

Historically, this system was used extensively with masonry bearing walls for early industrial buildings (from which it derives a description as *mill construction*). Typical details for the construction evolved from even more ancient usage, progressing to the increased use of steel connecting elements for column and wall framing. Some forms of these elements are shown in modern dress in Figure 6.3 and also in details for some of the buildings in Chapter 7 (Sections 7.3 and 7.9).

REFERENCES

Architectural Graphic Standards, 8th ed., Ramsey and Sleeper, New York: Wiley, 1988.

Fundamentals of Building Construction: Materials and Methods, 2nd ed., Edward Allen, New York: Wiley, 1990, chap. 4.

6.7.3 Miscellaneous Wood Systems

As discussed in Section 3.1, wood as a basic material can be used for a variety of products, from solid lumber to paper. Both framing members and decking for floor systems can be produced from solid, sawn wood, as illustrated in the preceding two sections. However, other products are also used. Some options for variations include the following:

1. Use of wood fiber decks; now commonly done for roofs.

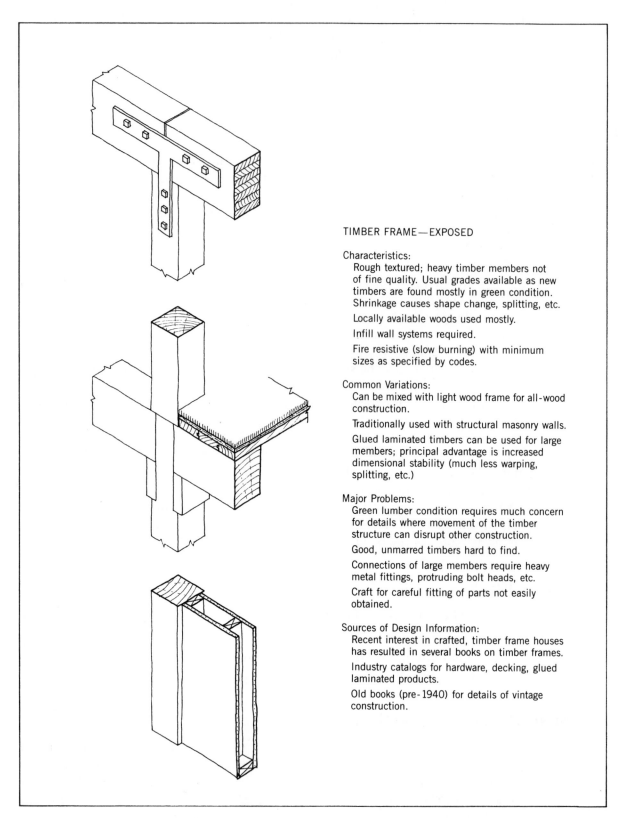

TIMBER FRAME—EXPOSED

Characteristics:
    Rough textured; heavy timber members not of fine quality. Usual grades available as new timbers are found mostly in green condition. Shrinkage causes shape change, splitting, etc.

    Locally available woods used mostly.

    Infill wall systems required.

    Fire resistive (slow burning) with minimum sizes as specified by codes.

Common Variations:
    Can be mixed with light wood frame for all-wood construction.

    Traditionally used with structural masonry walls.

    Glued laminated timbers can be used for large members; principal advantage is increased dimensional stability (much less warping, splitting, etc.)

Major Problems:
    Green lumber condition requires much concern for details where movement of the timber structure can disrupt other construction.

    Good, unmarred timbers hard to find.

    Connections of large members require heavy metal fittings, protruding bolt heads, etc.

    Craft for careful fitting of parts not easily obtained.

Sources of Design Information:
    Recent interest in crafted, timber frame houses has resulted in several books on timber frames.

    Industry catalogs for hardware, decking, glued laminated products.

    Old books (pre-1940) for details of vintage construction.

Figure 6.3   The heavy timber frame system.

2. Use of I-shaped or box beams with solid-sawn lumber flanges and plywood or fiber webs for spans greater than those feasible with solid-sawn lumber joists.
3. Use of light, prefabricated trusses with wood chords and steel web members.
4. Use of glued laminated beams as replacement for solid timbers for very heavy loads or long spans.

Products used for these, and other, variations are often priority items, produced by a single manufacturer, and possibly marketed only on a limited regional basis. They may be quite extensively used, but only in a limited area.

6.7.4 Light Steel—Open Web Joists

A light spanning system of long duration is that using closely spaced, light steel trusses, called open-web joists. These generally receive more use for roof structures, where the efficiency of the lightweight structure is more likely to be advantageous; roof live loads are generally lighter than floor live loads, spans often longer, and deflection and bounciness less critical.

Usage for floors usually involves slightly heavier trusses, closer spacing, and use with heavier decks. Still, the system may be effective, especially if the trusses are made to support a ceiling by direct attachment to their bottom chords. The latter is possible, while still retaining reasonable two-way use of the interstitial space, due to the openness of the truss profile—versus that of solid-web beams.

Figure 6.4 presents some details and characteristics of the open-web joist system, as utilized for floor structures. The general system can be used with various supporting structures, including steel W-section frames, as shown in Figure 6.4, but also including bearing walls of concrete or masonry.

An all-truss spanning system can also be used and is most effective in retaining the full two-way use of the interstitial space or in achieving relatively long spans. Figure 6.5 shows such a system, as developed for a low-rise office building.

A special application of the open web system is one using a wood deck; the deck being nailed to wood strips that are fastened to the tops of the steel joists. This is generally competitive with the system using composite, wood+steel joists. The choice of these systems is pretty much a matter of local marketing and availability from suppliers.

INFORMATION SOURCE

Steel Joist Institute

6.7.5 Heavy Steel—W Sections

Standard, hot-rolled W sections (formerly known as wide-flange sections), supplied in a wide range of sizes, are used to produce steel skeleton frames for many buildings—both those of modest size and those of record-setting height (see Figure 6.6). With the basic vertical column system, column-line beams, and possibly some trussing for lateral bracing all achieved with large steel members, it is quite common to use a general horizontal-spanning floor system that also uses W sections. Details and characteristics for a typical system of this type are shown in Figure 6.7.

The most common decking system in this case is that using formed sheet steel

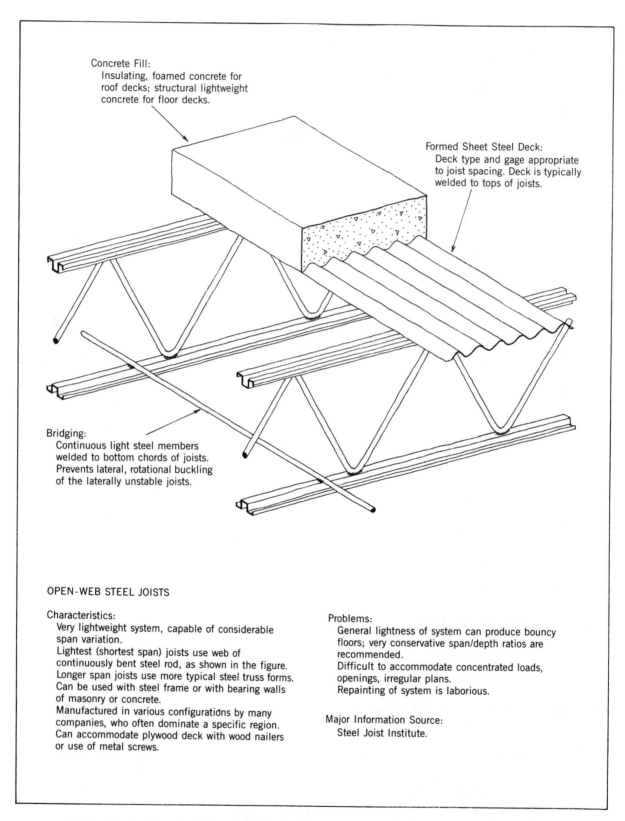

Concrete Fill:
Insulating, foamed concrete for roof decks; structural lightweight concrete for floor decks.

Formed Sheet Steel Deck:
Deck type and gage appropriate to joist spacing. Deck is typically welded to tops of joists.

Bridging:
Continuous light steel members welded to bottom chords of joists. Prevents lateral, rotational buckling of the laterally unstable joists.

OPEN-WEB STEEL JOISTS

Characteristics:
Very lightweight system, capable of considerable span variation.
Lightest (shortest span) joists use web of continuously bent steel rod, as shown in the figure.
Longer span joists use more typical steel truss forms.
Can be used with steel frame or with bearing walls of masonry or concrete.
Manufactured in various configurations by many companies, who often dominate a specific region.
Can accommodate plywood deck with wood nailers or use of metal screws.

Problems:
General lightness of system can produce bouncy floors; very conservative span/depth ratios are recommended.
Difficult to accommodate concentrated loads, openings, irregular plans.
Repainting of system is laborious.

Major Information Source:
Steel Joist Institute.

Figure 6.4   Floor framing with light steel trusses, produced as priority open-web joists by various manufacturers.

Figure 6.5   Utilization of steel trusses for roof and floor systems. Combination of open-web joists and supporting joist girders.

units with a concrete topping, as discussed in Section 4.1.2. Whereas light steel trusses may be quite closely spaced for floor systems, W sections are commonly spaced farther apart. Logical spacing has usually to do with the clear span to be achieved, as well as the type of deck and its span limits. Spacing is commonly some modular fraction of the center-to-center distance between supporting beams on the column lines.

Details for a specific application of this system are shown for building 6 in Chapter 7. As shown there and in Figure 6.7, it is common to suspend ceilings from the deck, partly because beams are usually too widely spaced to use for direct support. However, using the deck to hang the ceiling also frees the modular planning of the ceiling from that of the steel frame, which is often an advantage.

A major detail for this, and for all steel structures for multistory buildings, is the need to protect the steel to develop necessary fire resistances. The concrete fill above—or an actual concrete deck—may achieve protection from above, but the steel beams and the underside of the steel deck (if one is used) must also be protected. Underside protection can be achieved with a fire-resistive ceiling (as it usually is for truss systems), but a mineral-fiber, sprayed-on coating, as shown in Figure 6.7, is usually used for systems with W sections. Various details must be developed to allow for the fireproofing materials around beams and columns, affecting the minimum finished sizes of columns in plan and the real available dimensions of the interstitial space in the floor/ceiling system, for example.

The all-steel frame with W sections is extremely adaptable to both building size and form. Modifications may include the use of other steel elements, such as pipe columns, hanger rods, angle elements for bracing, and so on. Very large or special structural members can be built up from standard rolled products to create just

(a)

(b)

Figure 6.6 Building structures utilizing rolled steel sections: (a) Modest-size elements for a low-rise building. (b) Heavy elements for a high-rise building.

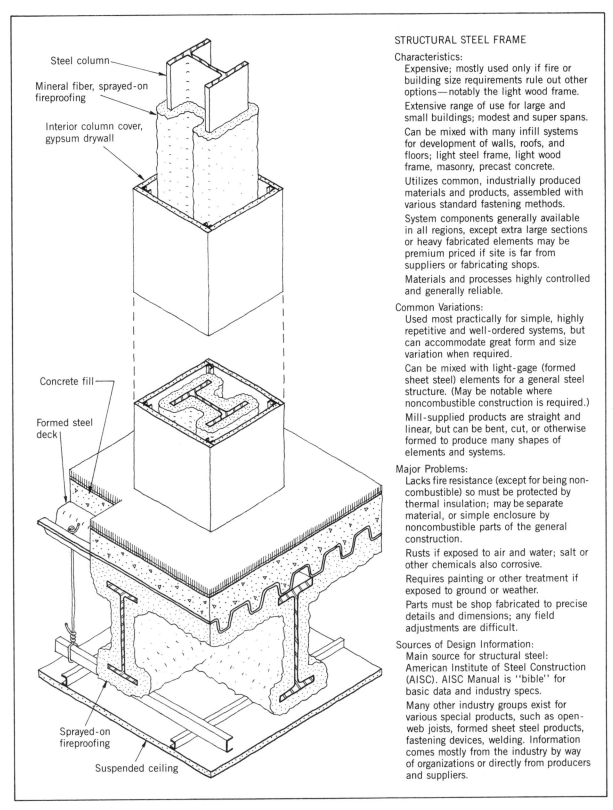

STRUCTURAL STEEL FRAME

Characteristics:

Expensive; mostly used only if fire or building size requirements rule out other options—notably the light wood frame.

Extensive range of use for large and small buildings; modest and super spans.

Can be mixed with many infill systems for development of walls, roofs, and floors; light steel frame, light wood frame, masonry, precast concrete.

Utilizes common, industrially produced materials and products, assembled with various standard fastening methods.

System components generally available in all regions, except extra large sections or heavy fabricated elements may be premium priced if site is far from suppliers or fabricating shops.

Materials and processes highly controlled and generally reliable.

Common Variations:

Used most practically for simple, highly repetitive and well-ordered systems, but can accommodate great form and size variation when required.

Can be mixed with light-gage (formed sheet steel) elements for a general steel structure. (May be notable where noncombustible construction is required.)

Mill-supplied products are straight and linear, but can be bent, cut, or otherwise formed to produce many shapes of elements and systems.

Major Problems:

Lacks fire resistance (except for being non-combustible) so must be protected by thermal insulation; may be separate material, or simple enclosure by noncombustible parts of the general construction.

Rusts if exposed to air and water; salt or other chemicals also corrosive.

Requires painting or other treatment if exposed to ground or weather.

Parts must be shop fabricated to precise details and dimensions; any field adjustments are difficult.

Sources of Design Information:

Main source for structural steel: American Institute of Steel Construction (AISC). AISC Manual is ''bible'' for basic data and industry specs.

Many other industry groups exist for various special products, such as open-web joists, formed sheet steel products, fastening devices, welding. Information comes mostly from the industry by way of organizations or directly from producers and suppliers.

Labels on figure:
Steel column
Mineral fiber, sprayed-on fireproofing
Interior column cover, gypsum drywall
Concrete fill
Formed steel deck
Sprayed-on fireproofing
Suspended ceiling

Figure 6.7   Interior systems with rolled W sections.

about any size or form required. The normally straight stock can also be bent or curved to produce other than rectilinear frames.

For floor beams, a common practice is to slightly curve the members for long spans, installing the beams with a slightly upward curvature of an amount estimated to be equal to the dead load deflection (from the weight of the construction). Thus the as-built floor will attain a dead flat condition and deflections will be due to live loads only.

INFORMATION SOURCE

American Institute of Steel Construction

6.7.6 Sitecast Concrete

Concrete offers exceptional resistance to fire, weather, wear, and many conditions that adversely affect other materials, causing rusting of steel, rotting of wood, and so on. It is also highly adaptive to various forms, having essentially no natural form. Where its advantages can be fully utilized, it can be used for floor structures for buildings with several different basic structural forms.

Figure 6.8 presents some details and characteristics of a basic sitecast concrete framing system, described as a slab-and-beam system. This system is essentially similar in nature to the heavy timber structure in wood and the heavy frame with W sections in steel. As in those systems, beams are usually quite widely spaced. While one limit for beam spacing is the thickness and resulting span capability of the concrete slab, this is usually not the actual limiting condition. To preserve their full fire-resistive character, slabs are usually a minimum of about 5 in. thick; as one-way spanning decks, they thus have a span capability of at least 12 ft, which is quite a wide spacing for beams.

The slab-and-beam floor system is the most adaptable and most widely used sitecast system. Other basic systems are shown in Figure 4.7, each having particular characteristics and applications. Some of the attributes and limitations of the systems are described in Section 4.1.2.

Concrete structures are used for various utilitarian purposes where they are economically competitive with other systems and where their particular properties are significant. There are also, however, some forms that can only be produced in concrete because of its nature as a cast material. One of these is the waffle system, which is essentially a two-way spanning joist system. This offers a very rich texture and a modular ordering system as an exposed ceiling, as shown in Figure 6.9.

The two-way solid slab offers an uncluttered underside, which may offer some advantage where elements occupying the interstitial space need two-way freedom, or where clear height and floor-to-floor height is restricted, as in underground parking structures. The flat plate is the ultimate thin floor structure, generally having no beams and a slab of only 8 to 10 in., as used for apartments, motels, dormitories, and so on.

Sitecast concrete structures are usually quite slow to erect and typically quite expensive. Their cost needs to be justified by some of the advantages to be gained from their use. Lack of need of protection for weather, fire, or wear can sometimes be cost-effective by eliminating need for additional materials to achieve these attributes.

Concrete floor structures are mostly used with building structures that are all concrete. However, as described in Section 6.7.8, they may also be used with masonry or steel elements in some cases.

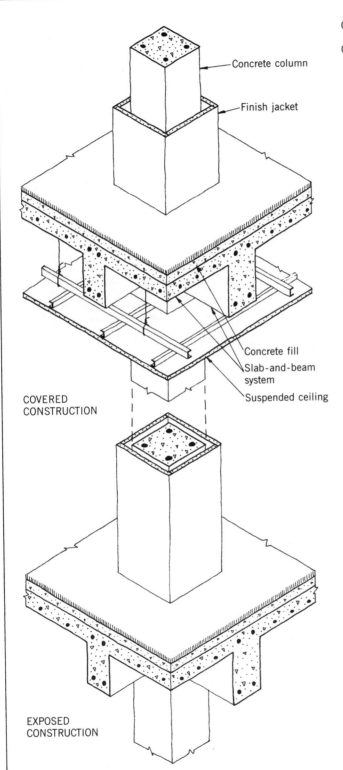

CAST-IN-PLACE (SITECAST) CONCRETE FRAME

Characteristics:

Expensive, especially for forming, finishes, reinforcing, and complex forms.

Requires utilization of major advantages of inherent properties—durability, stiffness, fire resistance, general solid feel.

Common construction, but requires major controls in design, specifications, details, and inspection if a quality of exposed work is expected.

Facilitates dissymmetry and wide variation of form within practical limits.

Produces natural rigid frame for stability against lateral forces (wind or earthquake) and a highly stiff and solid construction for ordinary gravity loads.

Lack of void space inside members creates problems for wiring, piping, etc.

Common Variations:

Type of horizontal structure: slab and beam, joist, waffle (two-way joist), two-way slab, flat slab, prestressed slab. System choice depends on loads, span, degree of continuity, regularity of bays, exposed details, required fire rating, etc.

May use concrete walls for bearing or shear walls.

Concrete construction may be fully or partly exposed or completely encased by finish materials.

Construction may be partly precast, using sitecast concrete to fill and tie together the structure.

Major Problems:

Facilitating piping, wiring, ducts, and built-in fixtures. Passage, attachment, and enclosure not as easy as with wood or steel frames.

Roughness and inaccuracy of construction: should be understood in detailing—with special concern for exposed structures.

Special detail problems: cold joints (at end of day's pour), color change in separate batches of mixed concrete, exposure of form ties, bar chairs, and other devices.

Remodeling and repair difficult.

Sources of Design Information:

Major industry sources:

American Concrete Institute (ACI).
Portland Cement Association (PCA).
Concrete Reinforcing Steel Institute (CRSI).

Principal design references:

CRSI Handbook for tabulated structural designs.
ACI Code for various requirements.
ACI Detailing Manual for standard details.

Figure 6.8   Sitecast concrete slab-and-beam system.

Figure 6.9   Sitecast concrete "waffle" system as an exposed structure. Provides strong ordering system for modules of curtain wall, lighting, and interior partitioning.

INFORMATION SOURCES

American Concrete Institute

Portland Cement Association

6.7.7 Precast Concrete

Precasting refers to the process in which concrete elements are cast somewhere other than where they are needed, and then must be moved into place. Casting may occur on the site, as in the case of tilt-up walls, which are cast on floor slabs and then moved into place by tilting them up and placing them with moving cranes. Precast elements used for floor structures, however, are usually factory cast and consist of standard products of some local casting facility. Common forms widely used are those shown in Figure 4.8.

Precast concrete spanning elements are usually also prestressed and have considerable potential for achieving long spans. Single or double tee forms can span great distances, but are used mostly for roof structures. Plank-form slab units are used for more modest spans for both roofs and floors, frequently with bearing supports of steel beams or masonry walls.

6.7.8 Mixed Floor Systems

Complete structural systems are often assembled from a single material, but can also occur with mixtures of different materials. Although plywood decks are used mostly over wood framing, they can be used with steel joists or beams, with nailing strips fastened to the top of the steel framing. Concrete decks are also used with steel beams, as sitecast slabs or as precast plank units.

A real composite system is that shown in Figure 6.1, consisting of a column and beam frame of steel W sections, joists of wood and plywood, plywood deck, and—most likely—a concrete fill on top of the deck. This is a real conglomeration of materials, but is quite practical and widely used for low-rise office buildings.

6.8 INTERIOR WALL SYSTEMS

Interior wall systems employ a wide variety of materials for finishes, but only a few basic forms of construction. Wall structures are discussed in Section 4.3, and general types of surfacings and finish materials are discussed in Section 4.4. This section considers some of the most common basic forms of construction for interior walls.

A major distinction to be made for consideration of wall systems is whether or not the wall does major structural tasks, principally for supporting vertical loads (bearing wall) or lateral bracing (shear wall). For the bearing wall this relates primarily to the basic wall structure, but for framed shear walls the surfacing is also of structural concern.

6.8.1 Wood Stud Walls

Figure 6.10 shows some typical details and summarizes some characteristics for the basic light wood frame wall with closely spaced studs. Where fire requirements permit its use, this is pretty much the definitive interior wall—for light bearing walls, shear walls, and nonstructural partitions.

The all-purpose surfacing for interior wood stud walls is gypsum drywall: a sandwich of two outside layers of paper and a core of gypsum plaster. This has largely replaced the wet plastered wall, plaster now being used mostly as stucco on building exteriors. The drywall surface must be finished, first by patching and smoothing the joints, then by applying some finish materials—the least being a coat of paint. Many other finishes are also applied as veneers or coatings over a backup support consisting of drywall, including solid wood paneling, ceramic tile, mirrored glass, polished marble, wallpaper, fabric, and carpet.

The hollow wall space presents a fire hazard that must be controlled with blocking (see discussion in Section 5.1.1.). However, the hollow space is mostly an asset, permitting the housing (and hiding) of wiring, piping, ducts, electrical outlets, switches, and light fixtures, recessed cabinets, and so on. This multiuse space is most appreciated when it is not available—as when solid walls of masonry or concrete are used.

Within limits, the hollow stud space can also be used to contain building columns of wood or steel, permitting an uncluttered interior space.

6.8.2 Steel Stud Walls

Figure 6.11 presents some details and characteristics of hollow wall construction achieved with steel framing of cold-formed, light-gage (formed sheet steel) elements.

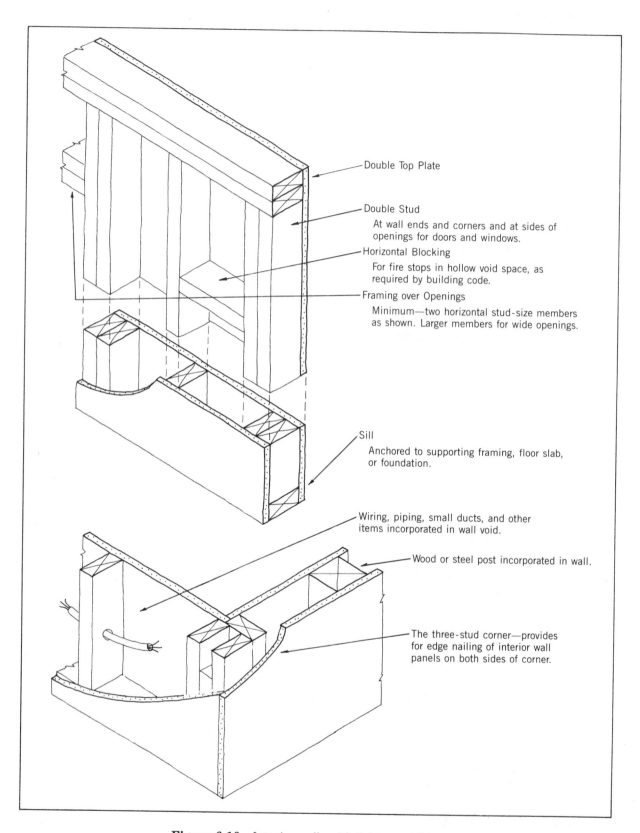

Double Top Plate

Double Stud

At wall ends and corners and at sides of openings for doors and windows.

Horizontal Blocking

For fire stops in hollow void space, as required by building code.

Framing over Openings

Minimum—two horizontal stud-size members as shown. Larger members for wide openings.

Sill

Anchored to supporting framing, floor slab, or foundation.

Wiring, piping, small ducts, and other items incorporated in wall void.

Wood or steel post incorporated in wall.

The three-stud corner—provides for edge nailing of interior wall panels on both sides of corner.

Figure 6.10   Interior walls with light wood frames.

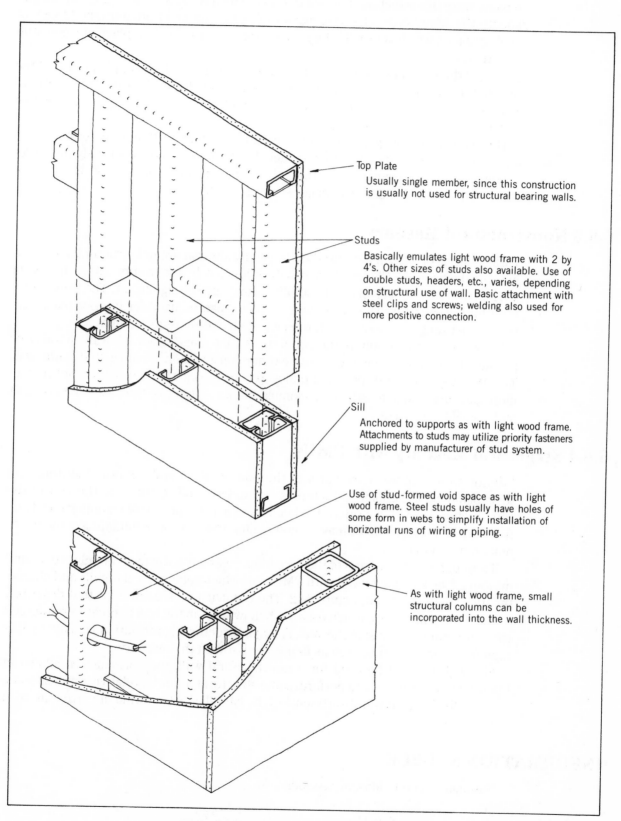

Top Plate

Usually single member, since this construction is usually not used for structural bearing walls.

Studs

Basically emulates light wood frame with 2 by 4's. Other sizes of studs also available. Use of double studs, headers, etc., varies, depending on structural use of wall. Basic attachment with steel clips and screws; welding also used for more positive connection.

Sill

Anchored to supports as with light wood frame. Attachments to studs may utilize priority fasteners supplied by manufacturer of stud system.

Use of stud-formed void space as with light wood frame. Steel studs usually have holes of some form in webs to simplify installation of horizontal runs of wiring or piping.

As with light wood frame, small structural columns can be incorporated into the wall thickness.

Figure 6.11   Interior walls with steel studs.

In many ways this structure imitates that made with wood and can generally accommodate the same range of surfacings and finishes. Its chief attribute is its lack of combustibility, which is usually the primary reason for using it in place of wood stud construction.

Most of the general attributes of the hollow wood framed wall apply here; add the fact that the steel studs are usually fabricated with holes in their webs (see Figure 6.11) so that accommodating horizontal runs of wiring and piping is one degree easier.

These walls are mostly used for nonstructural interior partitions or exterior curtain walls. This has to do mostly with the fact that they are used extensively for larger buildings, where a primary structure of other elements (masonry walls, steel or concrete frames, etc.) does the major structural tasks.

6.8.3 Nonstructural Masonry

Various forms of masonry are used for interior walls that are not structural (supporting only themselves). Choice of the construction may be for its appearance, but other attributes—such as lack of combustibility—may also be significant. The same forms used for structural walls—even exterior ones—can be used for interior partitions, but a wider range is possible with the lack of structural requirements.

Where appearance only is the concern, the interior "masonry" wall can also be produced with a thin veneer over some other construction—even a stud frame wall. This may be a matter of personal philosophy or ethics with a designer, but it is an increasing usage, due mostly to economics, but also simply to the steady loss of craft and available materials.

6.8.4 Structural Masonry with CMUs

Interior structural walls are parts of the building's general structural system, and their planning, the choice of materials and details, and relations to the rest of the construction must be considered. While less expensive or time-consuming construction may be possible, there may be compelling reasons for maintaining some continuity of use and details.

If appropriate on the basis of the preceding comments, the most popular structural masonry wall is that achieved with CMUs, for which some typical details and characteristics are presented in Figure 6.12. The construction illustrated is that described as *reinforced masonry*, which uses the hollow voids in the wall to create a reinforced concrete framework inside the wall. A major structural contribution is made by this rigid frame on its own, as well as in interaction with the masonry.

Although it could be used for a nonstructural wall, the construction shown in Figure 6.12 is most likely to perform some structural tasks. Its relative stiffness alone makes it likely that it will absorb some of the lateral effects of wind or earthquakes on the building.

INFORMATION SOURCE

National Concrete Masonry Association

6.8.5 Sitecast Concrete Walls

As with real masonry (not veneered), solid walls of concrete are most likely to occur on the interior as parts of the general building structural system. The general

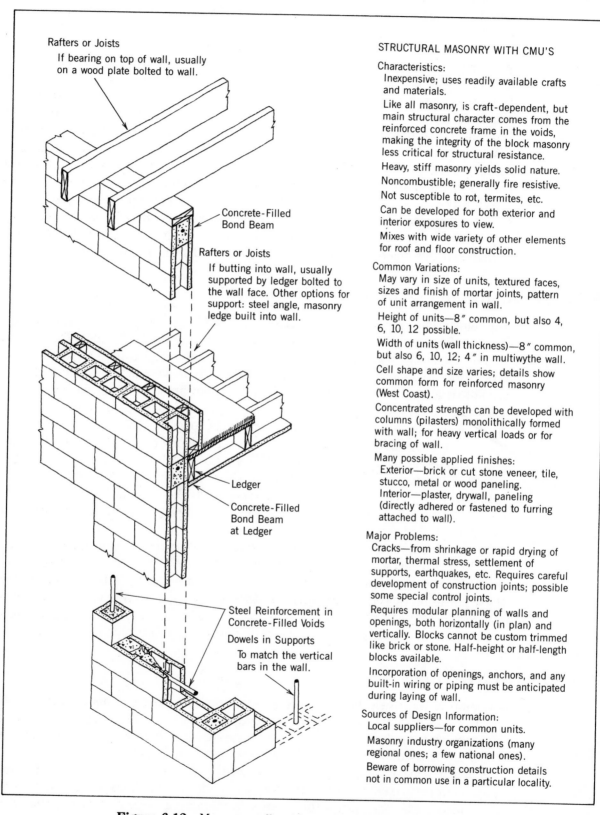

Rafters or Joists

If bearing on top of wall, usually on a wood plate bolted to wall.

Concrete-Filled Bond Beam

Rafters or Joists

If butting into wall, usually supported by ledger bolted to the wall face. Other options for support: steel angle, masonry ledge built into wall.

Ledger

Concrete-Filled Bond Beam at Ledger

Steel Reinforcement in Concrete-Filled Voids

Dowels in Supports

To match the vertical bars in the wall.

STRUCTURAL MASONRY WITH CMU'S

Characteristics:

Inexpensive; uses readily available crafts and materials.

Like all masonry, is craft-dependent, but main structural character comes from the reinforced concrete frame in the voids, making the integrity of the block masonry less critical for structural resistance.

Heavy, stiff masonry yields solid nature.

Noncombustible; generally fire resistive.

Not susceptible to rot, termites, etc.

Can be developed for both exterior and interior exposures to view.

Mixes with wide variety of other elements for roof and floor construction.

Common Variations:

May vary in size of units, textured faces, sizes and finish of mortar joints, pattern of unit arrangement in wall.

Height of units—8″ common, but also 4, 6, 10, 12 possible.

Width of units (wall thickness)—8″ common, but also 6, 10, 12; 4″ in multiwythe wall.

Cell shape and size varies; details show common form for reinforced masonry (West Coast).

Concentrated strength can be developed with columns (pilasters) monolithically formed with wall; for heavy vertical loads or for bracing of wall.

Many possible applied finishes:

Exterior—brick or cut stone veneer, tile, stucco, metal or wood paneling.
Interior—plaster, drywall, paneling (directly adhered or fastened to furring attached to wall).

Major Problems:

Cracks—from shrinkage or rapid drying of mortar, thermal stress, settlement of supports, earthquakes, etc. Requires careful development of construction joints; possible some special control joints.

Requires modular planning of walls and openings, both horizontally (in plan) and vertically. Blocks cannot be custom trimmed like brick or stone. Half-height or half-length blocks available.

Incorporation of openings, anchors, and any built-in wiring or piping must be anticipated during laying of wall.

Sources of Design Information:

Local suppliers—for common units.

Masonry industry organizations (many regional ones; a few national ones).

Beware of borrowing construction details not in common use in a particular locality.

Figure 6.12   Masonry walls with reinforced CMU construction.

development of wall shapes, reinforcement, and joints with other parts of the building structure must be developed as a combined structural and architectural design effort.

Concrete walls may be exposed to view or covered with a variety of materials. When covered up, a concern is for the means of attachment of finishes to the concrete surface, for which a wide range of hardware and techniques is available. Some fastenings are made with elements that must be cast into the concrete, which requires careful coordination during the construction. Easier to achieve are those fastenings that utilize adhesives, drilled-in anchors, or explosively driven anchors.

When exposed to view, the critical architectural concerns are for the general forming and finishing of the walls. Forming (casting of the concrete) entails the use of some forming units (plywood sheets, etc.), jointing techniques for the units, and typically the use of form ties that go through the wall and keep the two sides of the form from bowing out during placing of the wet, semifluid concrete. Ties are virtually impossible to hide on the wall surface; this is of no concern if applied finish is anticipated, but is something to deal with if exposed surfaces are used.

6.8.6 Demountable Partitions

For commercial interiors—most notably office buildings—it is common to rearrange interiors frequently, for new rental occupants or for new arrangements for existing renters. Structural walls and walls around permanent facilities such as stairs, elevators, restrooms, and duct shafts are quite difficult to move, but any other walls that are not structural or are not required for permanent fire control are fair game.

To make rearrangement of partition walls especially easy, many priority systems are available consisting of demountable wall units, together with various accessories for attachment to floors and ceilings, for mounting doors, for containment of wiring, and so on. Some systems are designed to interface with modular ceiling systems; some have associated office furnishings for shelving, counters, and so on.

Use of these systems is usually developed with the general design of interior layouts for individual building users and the overall design of furnishings, carpeting, wall finishes, and general interior decoration.

Large office interiors are sometimes designed with partial-height partitions; used together with office furniture, they define spaces and direct traffic, but leave the general space open at the ceiling level. While they are not walls in the usual sense, these partitions do serve many wall functions, including possibly the containment of some wiring, support of shelving, and so on. Units have standard modular dimensions, but can also be customized to some extent, particularly with regard to height.

6.9 MIXING SYSTEMS

Creation of the whole building interior requires the mixing of floor systems with interior supporting structure, ceilings, and the interior walls. There are common forms of mixtures, but just about anything goes, except that wood structures cannot generally be used to support steel, masonry, or concrete elements.

A major division generally occurs between parts of the construction that are reasonably permanent in nature and not easily subject to modification, and those that are relatively easy to modify without seriously disturbing the rest of the building. This may be a situation created deliberately—as in the case of office buildings built for speculative rental occupancies—or simply inherent in the construction due to the functional relationship of elements, for example, foundations and bearing walls (reasonably permanent) versus suspended ceilings and interior partitions (reasonably transient).

The whole building must eventually work as an integrated unit. This essentially means the satisfying of many individual relationships and issues; these may be considered individually, but must eventually be collectively synthesized into the whole design effort.

The remaining sections of this chapter present some of the critical coordinative issues regarding the design of the building interior.

6.9.1 Relations of Interior Walls to Floors and Roofs

Several different relationships can occur between the building walls and the horizontal structures for the roofs and floors. Major forms of relationship are the following:

1. Bearing walls: This means that the roofs and floors depend on the walls for support. In this case the plan location, structural capacity, and details of the connections between walls and the roof or floor structure become critical. Walls become essentially permanent and cannot be moved without major remodeling of the basic building structure. Also, for multistory buildings, walls in upper stories should be located either over walls below or over some major framing in the floor below to receive the loads from the bearing walls.

2. Nonbearing, but otherwise permanent walls: These are walls around permanent features, such as stairs, which must remain in place, but are basically partitions—that is, nonstructural. A problem here is that they must be constructed and connected to the building structure in a way that assures that they do *not* become structural. If built tightly against the underside of spanning beams, they can become unintended bearing walls; if made of rigid materials (masonry, real plaster, etc.) and tightly fit between columns or structural walls, they can become unintended shear walls. The trick is to provide the walls with the necessary vertical and lateral support they need, without making them unintended bearing or shear walls; not impossible, but requiring some study.

3. General partition walls: These are walls that serve to divide interior spaces, but do not surround any necessarily permanent features. The same problems occur here as with the nonstructural walls around permanent features—providing necessary support for the walls, but not making them unintended bearing or bracing elements. In addition, for deliberately intended speculative situations (rental offices, etc.), the wall construction should be of a type that is relatively easily removed and relocated or reconstructed.

A second relationship to be considered here is that between the ceiling and the walls. As shown in Figure 6.13, there are different relationships, depending on the general interior design, but also on certain functional situations. For structural walls, the top of the wall must be in contact with the underside of the supported roof or floor structure; thus the ceiling stops at the wall face, defined a room at a time. For nonstructural walls, the wall top may be continued to the underside of the overhead structure with the ceiling treated as for permanent walls. Depending on the form of the ceiling construction or on general building planning considerations, however, it may also be possible to have the nonstructural walls terminate at the ceiling level as shown in the lower part of Fig. 6.13. However, for fire separation or for security

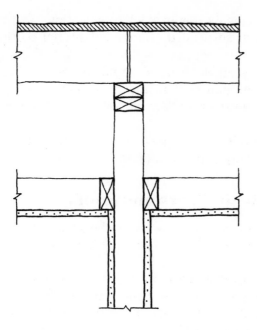

Bearing Wall

Wall supports upper spanning structure; studs continuous through ceiling level; ceiling joists supported by ledgers on stud faces.

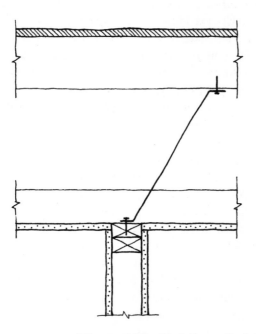

Nonstructural Wall

Studs to ceiling level only; top of wall braced from structure above; ceiling joists can sit on top of studs or be supported on ledgers if different ceiling heights on opposite sides of wall.

Figure 6.13   Variations of relationship between walls and ceilings.

between areas occupied by different building tenants, it would be desirable to have the situation shown in Figure 6.16b, even for intentionally removable walls.

6.9.2 Integration of Interior and Exterior Construction

It is possible to consider interior and exterior conditions separately in many regards; however, eventually the whole building must be dealt with. The inside surfaces of exterior walls become parts of the building's interior, for example, and their appearance, light-reflecting characteristics, flame spread, and other properties must be handled with the general interior design.

Windows are prominent features of the building exterior, but are also intimately related to individual interior spaces. They are also very hard to change, so they become generally fixed items for any form of future interior rearrangements. Future new interior partitions, intended to intersect the exterior walls, cannot do so in the middle of a window, unless the window is developed with vertical mullions that can facilitate such a connection.

Doors in exterior walls present situations similar to those just described, in addition to which they are often required fire exits and any interior rearrangements must retain their usefulness in this regard.

6.9.3 Stacking

Multistory buildings are like layer cakes; each layer may contain something different or be essentially the same. Floor structures become transitional elements and must be related to spaces above as well as to those below. Interior walls may be basically the same on each level, totally independently arranged on each level, or actually vertically continuous in the building—as for elevator shafts.

The general building form may change from level to level, incorporating profiles with pyramidal or inverted pyramidal forms, and—quite commonly—setbacks for a stepped-back ascending profile change. Exterior profiles may also remain unchanged while interior form changes—creating atriums, upper-level open spaces, and so on.

Development of the building interior, in particular the incorporation of major structural elements and vertically rising service elements (elevators, ducts, etc.), must be carefully done as a three-dimensional design activity. Plans tend to be worked out a level at a time, often enough of a problem at one sitting! However, the superimposing of plans and general total spatial development of the building must be kept in mind.

6.9.4 Infill

The idea has been presented in many preceding discussions that the building construction can be viewed as being divided between elements that are of various classifications. Major distinctions can be made between the following:

Structural versus nonstructural: That is, what is part of the overall, major structural system of the building, and what is self-structural only.

Permanent versus transient: That is, what must generally remain in place, and what can be removed, relocated, reconstructed, or otherwise changed without essentially disturbing the overall, continued functioning of the building.

Much of the building construction is nonstructural and potentially transient in nature. This does not classify it as unimportant in terms of architectural design, basic functional requirements, or essential user needs. It does relate, however, to the character of the construction and to the range of materials and systems that can be used.

Buildings are typically created by layering various infill onto a skeleton structure. Initial layers may themselves be structural, as when decking or wall sheathing is added to a frame. Additional layers or inserts achieve various effects: insulation, weather protection, finished appearance, service to power outlets, and so on. Eventually the whole, multilayered construction is developed.

When layers are peeled back—for refurbishing, remodeling, or repairs—the hierarchies and dependencies must be kept in mind. The *ability* to do remodeling, repairing, and so on, must sometimes be reasonably anticipated in developing the original construction. There is a tendency in design to treat the finished building as a static condition—suddenly, totally created and forever the same.

Buildings are typically very expensive, represent major investments, and anticipate a long life. Anything intelligent done in the design stages to provide for that long life will be appreciated later. Developing the general construction so that the realities of the hierarchical relationships can be recognized in future changes is a step in this direction.

6.10 INTEGRATION OF BUILDING SERVICES

Most buildings have electrical power, lighting, water supplies, waste piping, and HVAC systems. Multistory buildings have elevators, escalators, and vertical shafts for wiring, piping and ducts, chimneys, incinerators, mail chutes, and so on. Operation of the various services requires equipment that must be housed inside, on top of, or under the building.

Incorporation of service elements is a major chore in interior design, requiring considerable cooperation between the designers of the various separate systems. Overall coordination generally falls to the architectural designer and must be accomplished along with the design of the general interior construction. Some major considerations that must be made in this effort are the following.

6.10.1 Structural Interference

Structural interference relates to the problems of inserting all the nonstructural elements (including nonstructural construction) in a way that does not critically disturb the required functioning of the building structural system. Of typical concern are holes cut through walls or webs of beams, floor and roof openings that disturb orderly layouts of framing, and a need for horizontally continuous spaces that preclude locations for columns, bearing walls, shear walls, X-bracing, and so on.

The structure must function for safety of the building. If it cannot do so adequately and accommodate the other necessary building functions, something has to go. The structure is not inviolate, and some forms of compromise are to be expected. This is a two-way street, and a lot of judgment is required.

6.10.2 Fundamental Needs of Services

It helps if the designer understands some of the primary needs and limitations of the various building services. For example:

1. Waste drainage piping (from sinks, toilets, etc.) must facilitate gravity flow of liquids. This is especially a problem for long horizontal runs. A single pipe may theoretically fit inside some interstitial space, but its position must change if it has a long horizontal run.
2. Waste drains must be vented—generally through the roof and generally directly

vertically as much as possible. This is a problem for the roof, as well as all the construction between the vented fixture and the roof.

3. Hot-water pipes and heating ducts get hot; chilled-water pipes and air conditioning (cooled air) ducts get cold. Surrounding construction can be heated or cooled. Thermal expansion and contractions can create problems. Chilled objects can sweat and create moisture problems.

4. Electrical wiring must frequently be altered and various other elements should be accessible for adjustments, repairs, alterations, and so on. Embedding elements in the permanent construction can create many problems.

5. Equipment with moving parts—notably motors, fans, pumps, and chillers—create both noise and direct, physical vibrations that can be annoying. They should be isolated—possibly by good planning of their locations, but also by use of vibration isolators, sound-insulated partitions, or other special construction.

As noted previously, compromise is the name of the game. Ideally, all building systems should operate or exist with optimal conditions, but economic feasibility, technical limitations, and some priority of design values and goals must be recognized.

6.11 FACILITATION OF MODIFICATION

A peculiar requirement of modern buildings is their general need to facilitate modification. The dynamic nature of our society makes this imperative. People relocate regularly; businesses reorganize; the urban fabric continuously changes—growing, decaying, regrouping; and rapid changes in technology bring many sudden demands for facilitation of new opportunities.

Playing against this rapidly shifting scene is the great expense of building costs, which makes the writing off of initial construction over a long period mandatory for most investors. The net effect is that buildings must be built for permanence but somehow be easily, feasibly, and preferably rapidly modifiable.

Ease of modification is a design goal and involves both basic planning and the final decisions for choices of materials and details of the construction. It also needs to permeate the whole design effort, including that of the structure and all the building services where possible. This is relatively easy for some things, such as wiring, ceilings, and plumbing fixtures; not so easy for sewers, foundations, and shear walls.

Designing for special requirements, such as ease of modification, may require extra effort and extra cost of the construction. In some ways, however, it merely means picking between competing alternatives with a view to those that facilitate ease of change. Bolted joints can be undone, for example, while glued or welded ones cannot.

7

Case Studies

The purpose of this chapter is to present a number of examples of building types and a range of construction usages and to illustrate the development of the general construction for each example. The major intent is to present the construction in terms of whole systems for specific cases. Design of the whole system thus becomes the major concern, and the variability of the individual elements can be viewed in the context of their role in the general construction.

The presentations here are meant to compliment the materials in preceding chapters. Critical isolated issues are dealt with more extensively in those chapters, and that material should be used as a resource for the more detailed information on individual items shown in these examples.

However, building designers must ordinarily face the problem of the whole building, even though the actual design may be worked out by concentrating on one part at a time. Here, the presentations begin with consideration of the building's whole general construction system, followed by presentations of some of the primary increments of the construction with all of the significant parts displayed.

The building designs presented here are not meant as illustrations of superior architecture. The interest here is directed mostly to setting up reasonably realistic situations that cover a range of circumstances in regard to design of construction. It is not the intention to present boring, ugly buildings, but the use of a limited number of examples to display a broad range of construction use is the dominant concern.

An attempt has been made to show construction usage and the form of details that are reasonably correct. This is, however, a matter for much individual judgment and is often tempered by time and location. There are not many situations in which only one way is correct and all others are unequivocally wrong.

Reasonably correct alternatives are often possible, and the particular circumstances for an individual building can make a specific choice better on different occasions. Differences in climate, building codes, local markets, building experience, or community values can produce a way of doing things that is preferred. Over time, changes in any of those areas of influence will produce variety and, it is hoped, evolution toward a better way of doing things.

The concentration in this book is on interiors, but it is not really possible to consider interior construction fully independent of the construction of the building's general enclosure systems. Enclosure for these examples is considered in the book that was the first volume of this series—*Building Construction: Enclosure Systems.* The enclosure system (roofs, exterior walls, windows, and doors) is shown in sufficient form here to permit an understanding of the context for development of the building interior. Of particular form in this regard are the problems of window form, detail, and location and the development of the interior surfaces of the exterior walls.

Finally, although incorporation of building services is a major concern for the development of the building's interior construction, the full development of these systems cannot be shown in detail here. However, some critical issues that significantly affect individual elements of the interior construction are discussed in the examples in some situations.

7.1 BUILDING 1

Single-Family Residence
Light Wood Frame

This is a building of modest proportions; not impressive to architectural critics, but a castle by world housing standards. As a type of building—and in this general size and general character—it is probably the most widely constructed building in the United States. Cape Cod cottages used this basic structure hundreds of years ago and are still imitated as a style in extensive housing developments. The form of house shown here, in Figure 7.1, is that of a typical suburban split-level, but many details are really reminiscent of the basic Cape Cod cottage.

While the general style of construction may stick with us, the materials, products, and details are considerably changed. What does not change much over time or region is the basic structure and general form of the light wood frame system—using studs, joists, rafters, sheathing, and decking. The characteristics of this system, as applied to development of interior construction, are discussed in Sections 6.7.1 and 6.8.1.

Development of the building enclosure tends to be strongly affected by local conditions. This is less true for interiors, especially with regard to the influences of climate and favored exterior finishes. The same basic interior development can be essentially achieved within a variety of enclosures, as the several variations of models in large housing developments will illustrate. Viewed differently, it is possible to wrap a common interior with an endless variety of enclosures.

Basic Construction

The general form of construction is shown in the details in Figures 7.2 and 7.3. The light wood frame is still predominant in use for this type. The standard industry sizes for the 2 × 4 studs, the 2 × 6, 8, 10, or 12 joists and rafters, and the 4-ft by 8-ft panel size for sheathing and decking set a modular system of dimensioning in place. Frame spacings of 12, 16, 24, 32, 48, and 96 in. are obvious, and the 3.5-in.-wide void space in walls is common.

Some dimensioning off the basic modules is to be expected to achieve specific room dimensions, but the pressure to reduce waste makes use of the popular, readily available units of materials important. A major cost factor—here as for all building construction—is the need to reduce to a minimum the requirements for on-site labor, particularly those of the skilled crafts.

Because of the popularity of this construction, and in anticipation of its standard dimensions, various items such as windows, doors, built-in wall cabinets, and recessed electrical fixtures are produced specifically to fit the modular dimensions. These items are also typically factory-made in a relatively complete form—designed

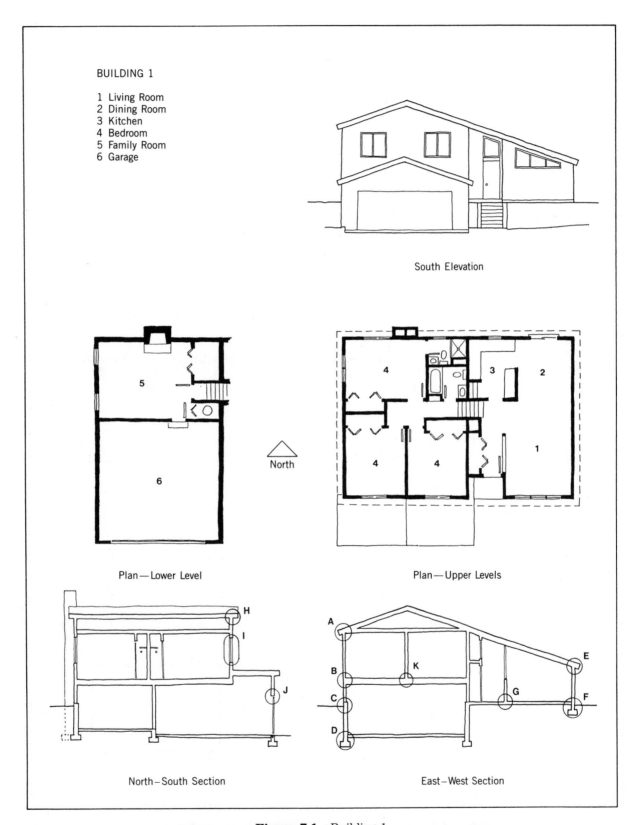

BUILDING 1

1 Living Room
2 Dining Room
3 Kitchen
4 Bedroom
5 Family Room
6 Garage

South Elevation

North

Plan—Lower Level

Plan—Upper Levels

North–South Section

East–West Section

Figure 7.1 Building 1.

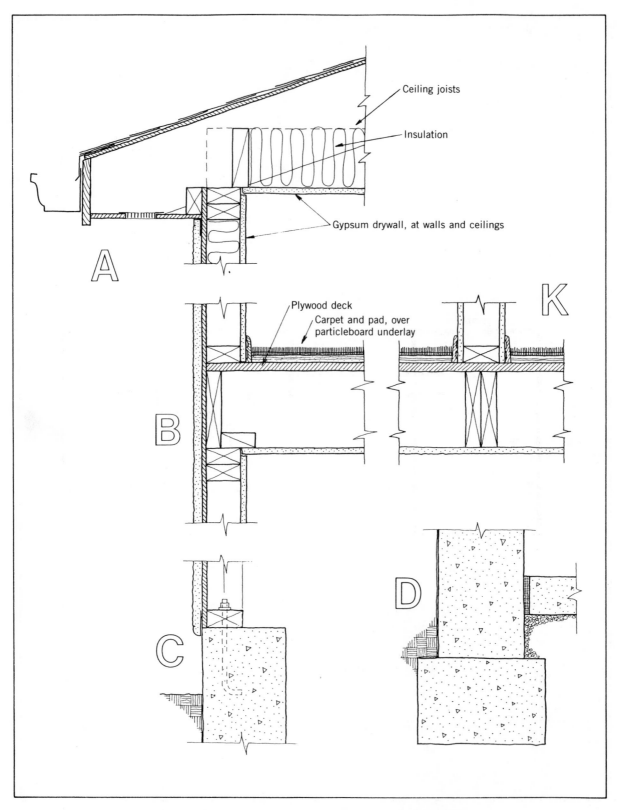

Figure 7.2   Building 1: general details of the building shell.

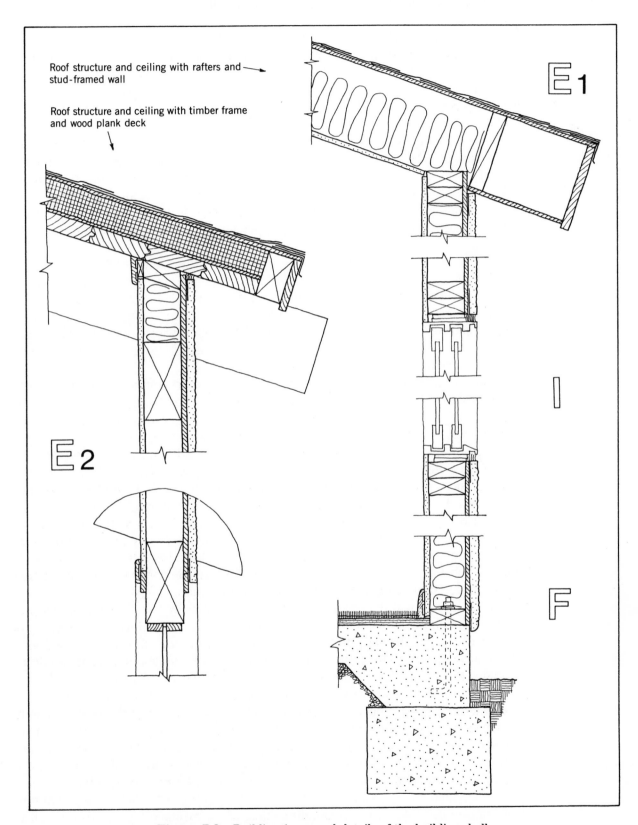

Figure 7.3 Building 1: general details of the building shell.

to permit their easy insertion into the basic structure with little need for site work to finish them.

However, despite the strong control of the modular dimensions of standard products, this structural system has immense potential for the accommodating of variations of building form, size, and details. The materials are relatively easy to work with—on or off the site—and custom forming can be done virtually without limit. This adds considerably to the enduring popularity of this building type with designers and builders.

There are, of course, other possible structures for this type of building. The light wood frame can be emulated in steel, using light-gage (formed sheet steel) elements. Use of steel offers the principal advantage of reducing the mass of combustible materials in the building. This may be a desirable feature in some situations, or an actual building or zoning requirement in others. It is hardly ever, however, an economic advantage; nor does it make the design or construction easier or faster to achieve.

The void spaces in walls, as well as those between rafters and joists, are utilized for many practical purposes. Primary ones are the containing of wiring, piping, ducts, and the various items for the systems for electrical power, lighting, water supply and waste plumbing fixtures, and miscellaneous elements for phone, intercom, thermostat, doorbell, and security alarm systems. The hollow wall space can also contain columns for strengthening of the frame in some cases, without having the frame intrude in the building occupied space.

While the 2 × 4 is the workhorse of the light wood framing system, and the 3.5-in.-wide stud space most common, there are sometimes compelling reasons for variation of wall thickness or different-size studs. For tight planning situations, nonstructural walls of thinner dimensions can be produced by using 2 × 3 or even 2 × 2 studs. More often, however, there is a need for thicker walls—or more specifically, for a wider void space. The two most common reasons for a wider void are the need for more insulation in exterior walls and the need to accommodate large items, such as air ducts or extensive plumbing. (See details in Figures 7.7 and 7.8.)

Taller walls may require larger studs, the general limit for 2 × 4 studs being about 14 ft. Excessive vertical loads or very high wind loading may also require more strength from the wall structure. For the building shown here, however, most walls will typically be made with 2 × 4 studs, except for possible needs for more insulation or accommodating of equipment, as described previously.

The 12-in. nominal depth (actually 11.25 in.) is usually the limit for rafters and joists, although the industry supplies 2 × 14 and 2 × 16 sizes. Horizontal-span limits are typically established by deflection criteria—limiting sag or bounciness of floors. For spans of more than about 20 ft for roofs or 16 ft for floors, it is advisable to consider the use of light trusses or fabricated, composite elements of lumber + plywood or lumber + fiberboard to achieve spanning elements of greater depth.

In the past, sheathing for walls and decking for roofs and floors was mostly achieved with solid sawn wood boards. Early in this century, these were gradually replaced, with decking mostly done with plywood and wall sheathing with plywood or various wood fiber products in panel form. Now compressed wood fiber products are steadily replacing most applications for plywood for decking and sheathing.

It is possible, of course, to use any form of structure for a house. Steel frames, masonry, reinforced concrete, aluminum skins, and fabric have been used—as well as ice, mud, twigs, and animal skins. The homemade or highly crafted custom-designed house offers endless possibilities and some notable, outstanding examples. For everyday consumption, however, the light wood frame still stands as the most widely used system.

Interior Construction and Finishes

From the point of view of construction, the development of the interior of this building is considerably predictable. Interior form, room arrangements, and finish materials can vary endlessly, but the basic construction and the general palette of materials is about as common as the 2 × 4 stud.

Development of the interior begins with the consideration of the overall construction as determined by the basic shell of the building enclosure. This presents the following:

Underside of the roof structure with sloping rafters and the plywood deck

Inside of the exterior walls with the studs and the exterior wall sheathing

Insides of below-grade walls; most likely, of sitecast concrete or CMU construction (concrete blocks)

Top of the concrete paving slabs

None of this construction is likely to be very desirable in an exposed condition, with the following possible exceptions. Concrete pavements may be acceptable in the basement and garage spaces, if the concrete surface is finished to a very smooth, hard quality. Insides of below-grade walls may be acceptable in the garage; less likely in the basement when the space is really occupied, not just for storage and equipment. The exposed stud construction may be acceptable in the garage.

Mostly, however, the basic structure needs to be covered for development of interior spaces. For the roof, a first question is whether the sloping form is acceptable for a ceiling. If so, the ceiling may be developed as shown for the living room in the section in Figure 7.1 and the details in Figure 7.2. The minimum construction in this case would be the application of drywall (as shown) or some other surfacing, such as wood paneling or lath and plaster. An alternative here would be to use the timber frame and plank deck as an exposed structure, as shown in detail E2 in Figure 7.3.

With the ceiling surfacing applied directly to the bottom of the rafters, the enclosed void space exists only between rafters. If ceiling-mounted lighting is used in this area, it may be necessary to run wiring perpendicular to the rafters; this is a common situation, but one that does require drilling holes through a number of rafters.

If an all-air, central HVAC system is used, it will be difficult to deliver air to the far side (east side) of the living room and dining room with the ceiling as shown. Ducts might be placed beneath the concrete pavement, but this is expensive construction. There are many options in this situation, but it is one consideration to be made in developing the interior design.

The high, sloping ceiling may be desirable in the living, entry, and dining areas, but questionable in the small kitchen—a trap for stale air, for one. In this case, a lower ceiling could be framed separately to the kitchen walls.

In the bedrooms in the upper level, it would also be possible to have the sloping ceiling, as in the living room. However, the sections in Figure 7.1 and the details show the use of a flat ceiling with an attic space above it. The roof and ceiling here could be framed with prefabricated trusses, with the top chords forming the roof and the bottom chords the ceiling. The same trusses could also be used in the front portion of the garage.

A major structural problem here is the achieving of a clear span across the garage. This is easy for the roofed-over portion in the front part of the garage, but the rear part is beneath the bedrooms, as shown in Figure 7.1. A possible solution would be to use a large beam beneath the wall at the front of the bedrooms, spanning across the garage to posts in the garage walls. Then relatively short floor joists could be used between this beam and the back wall of the garage.

The garage is shown with its floor at some distance below the room behind it. This may be developed in the site design, but also relates to a code requirement to have the floor lower so that gas fumes that lie at floor level do not creep into the occupied space through the door. Code requirements must also be satisfied for fire separation between the garage and the rest of the house, affecting the details of wall and floor construction and choice of the door materials.

Except for the baths, kitchen, and front entry area, the most commonly used floor finish these days is carpet, although many other options are possible. For the construction, the only concern is the anticipation of required thickness allowances. If thick tiles or solid wood flooring is used, the structure must be sufficiently depressed in these areas.

A problem that sometimes occurs is where flooring materials change in the same level. For example, suppose that carpet is used in the living and dining rooms, vinyl tile or linoleum in the kitchen, and thick ceramic tile in the entryway. Keeping the wearing surfaces of these materials at the same level means the accommodating three different thicknesses of flooring materials. Not to complicate the installation of the concrete pavement or preclude possible future changes of flooring, the top of the concrete should be set at the lowest level to accommodate the thickest anticipated materials. Then fillers can be used to build up to the necessary level for thinner materials. Details for this should follow recommendations of the suppliers of the flooring materials.

Special Interior Construction

The details indicate very ordinary solutions for the construction of the major portions of the house. There are also a number of special areas that must be considered, as well as various options for their construction.

Stairs.

Stairs may be built with ordinary wood framing and wood steps, with the necessary provisions of openings in the floor framing and through the walls as required. The steps to the lower level and to the garage, however, also sit on the concrete structure, and some options are possible for their construction of concrete, in conjunction with the general construction of the foundations, basement walls, and floor pavements. Possible details of this construction, showing some options, are given in Figure 7.4.

Fireplace.

The fireplace, as shown in Figure 7.5, is developed as an all-masonry structure in a very traditional manner. The chimney accommodates two flues—one for the fireplace inside the house and one for the barbecue on the rear terrace. While this

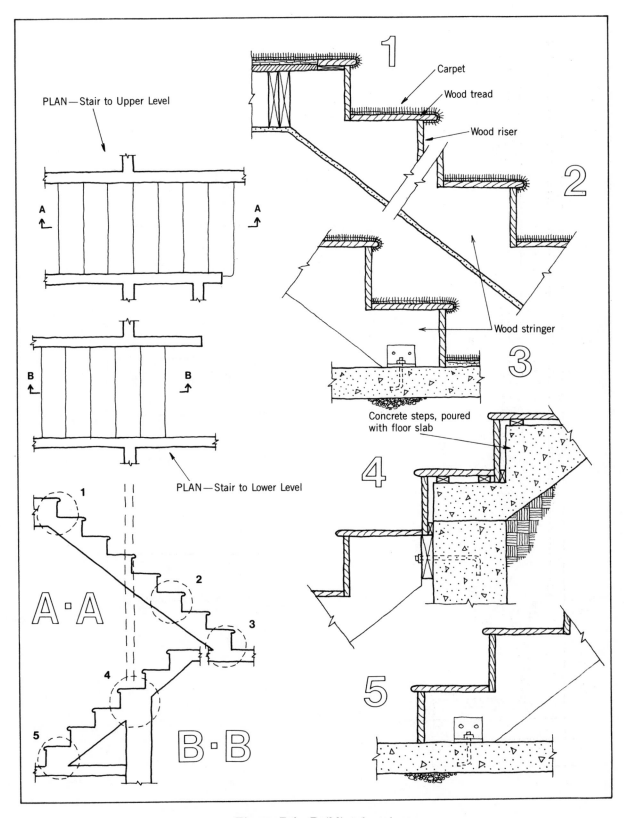

PLAN—Stair to Upper Level

PLAN—Stair to Lower Level

Carpet

Wood tread

Wood riser

Wood stringer

Concrete steps, poured
with floor slab

A·A

B·B

Figure 7.4   Building 1: stairs.

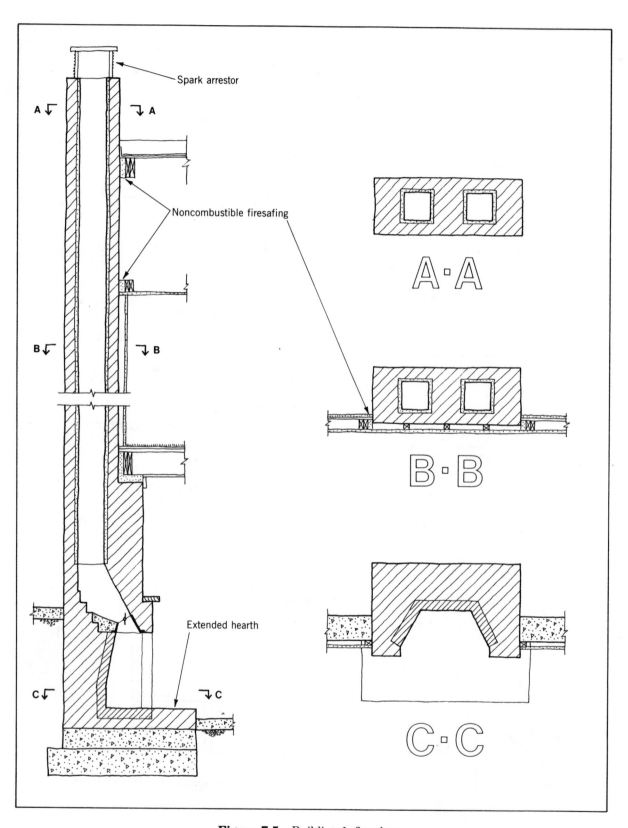

Figure 7.5   Building 1: fireplace.

construction is still possible, it is very expensive, and prefabricated units are now available to achieve the fireplace itself and the chimney. If the prefabricated elements are used, a masonry shell could still be built around them, but would be much simpler in construction.

Kitchen.

Although the image of the kitchen as a major activity center is fading (what with eating out, microwave ovens, two-career families, etc.), it is still likely that some basic elements would be provided here. These may include provisions for built-in cabinets and countertop units, built-in ovens and ranges, sinks, and an exhaust fan. The construction may accommodate these in various ways, but probably with little basic alteration of the system. In this building, the locations of the kitchen, baths, and lower storage room offer some possibilities that are developed in the details in Figure 7.6. This is all very tight planning and requires careful study for simplification of the construction and economizing of space.

Bathrooms.

Figure 7.7 shows some details for the construction at the bathrooms in the upper level. This is a location of considerable plumbing, for three different services: cold-water supply, hot-water supply, and wastewater (sewer) piping. Installation of the piping and the fixtures requires various considerations for the plumbing, including the following:

Space within the voids of walls and floors for the pipes and some portions of recessed fixtures

Allowance for drainage of horizontal piping runs, especially for the sewer lines

Provision for extension of vent piping through the roof, with vents for each drain for sinks, shower, bathtub, and toilets

Cutouts as required in the floor deck for the toilets and the drains for the shower and bathtub.

In order to help accommodate the concentrations of vertical pipes, the walls between the two bathrooms and between the bathrooms and the kitchen are thickened slightly. This can be done by using larger studs (2 × 8?) or by using two rows of studs. The wall between the lower and upper portions at this location is quite chopped up with the plumbing, the stairs, and some provisions for the HVAC and hot-water services. This would require some careful study, once the exact form of the equipment is determined.

Divider at Entry.

This is an element that achieves some separation between these two areas. Due to the rather long span of the roof from the outside wall of the living room to the wall between the levels, this divider may also be used to support the rafters. This structural function would call for a somewhat sturdier form of construction; however, the divider is quite tall and would need some substantial stiffness in any event. Some possible details for the divider are shown in Figure 7.8. These assume that the divider

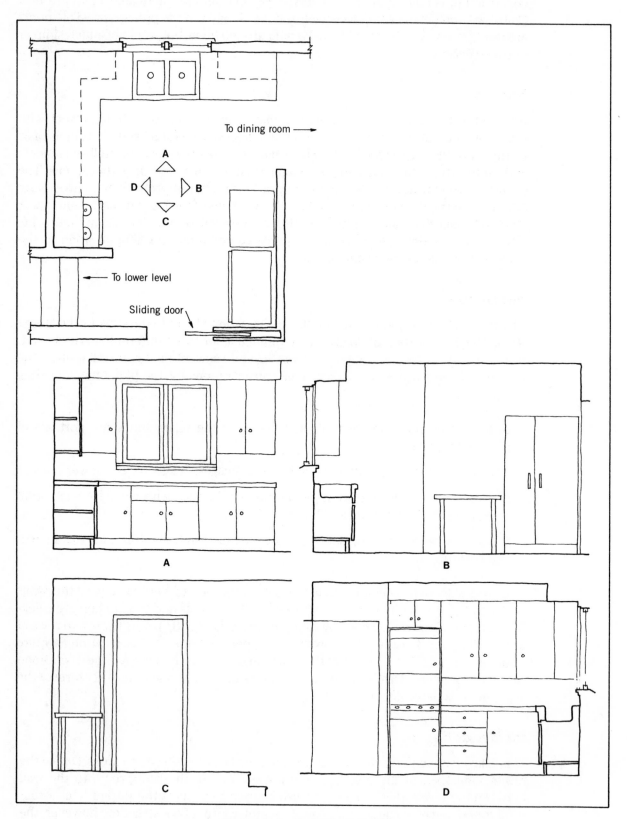

To dining room →

A

D ◁   ▷ B

C

← To lower level

Sliding door

A

B

C

D

Figure 7.6   Building 1: kitchen.

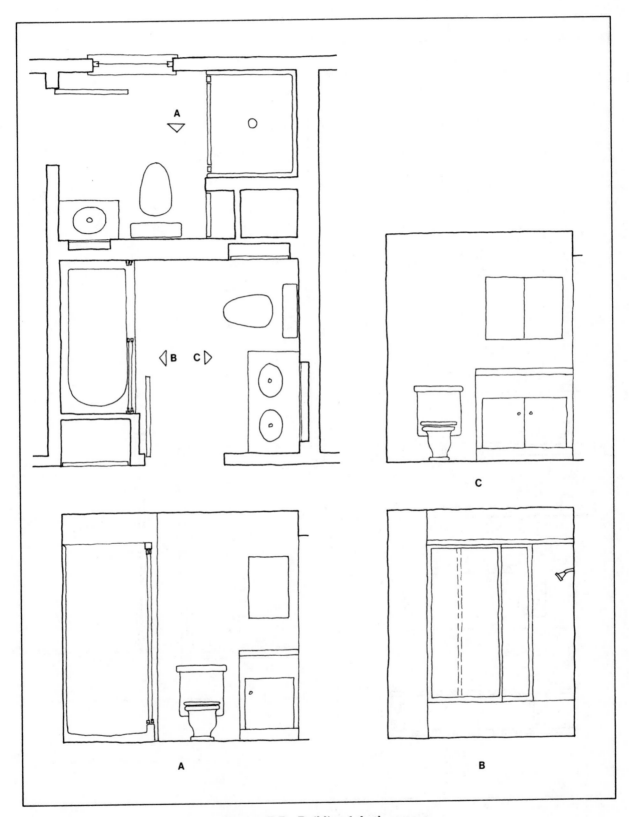

Figure 7.7   Building 1: bathrooms.

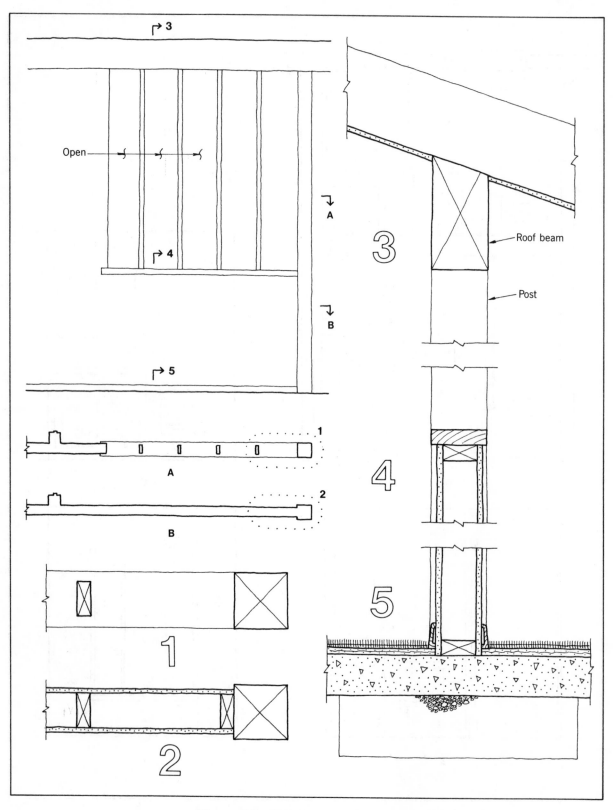

Figure 7.8   Building 1: divider at entry.

supports the roof. The low, solid portion of the divider wall allows some furniture in the living room to be placed along the wall. Many other details are possible for this semistructural, semidecorative element.

Customization

Despite its extremely common construction, there is considerable opportunity for "customizing" the interior design of this building. This is most easily accomplished by choices for finish materials, but some minor variations in the construction itself can also be made to further modify the form of the interior. Most notably, interior walls may be rearranged, as long as their structural functions or use to house service elements is adequately recognized.

Many special enhancements can also improve various aspects of the building. Sound treatments of some portions, such as walls between bedrooms or floors separating levels, may increase privacy for the occupants. Extra insulation in outside walls will be significant for cold climates. Special efforts to dampproof the basement walls and floors will be important to the lower-level occupied space.

7.2 BUILDING 2

One-Story Commercial Light Wood Frame

This is a small, one-story building with a flat roof and a parapet wall on all sides (see Figure 7.9). Although the building plan as shown is generally symmetrical about a central corridor, the building exterior is developed to relate to a site orientation that results in one long side being treated as the front (street-facing) and the opposite side as the rear. A short cantilevered canopy is provided on the front side, and the roof surface is sloped to the rear. The most likely method for draining the roof surface in this situation would be by scuppers through the rear parapet and vertical leaders on the rear wall.

The general details of the construction are shown in Figure 7.10. These reveal the structure to be a light wood frame with structural plywood siding on the rear and ends and brick veneer on the front. The floor is a concrete pavement and the roof structure consists of a plywood deck on closely spaced, clear-span trusses.

Use of the clear-span roof structure permits interior walls to be placed at will, with remodeling for other desired arrangements in the future easily achieved. This form of structure is common for commercial buildings built primarily as speculative rental properties, as it allows for a wide range of prospective tenants.

Despite the accommodation of interior rearrangements, certain features become relatively fixed and not subject to easy change. Primary elements in this regard are the building foundations and major structure. Also difficult to change are windows and doors, making some reasonably accommodating locations the most desirable. Plumbing—particularly sewer lines built into the pavement—are also hard to change.

Changes will be made easier if most services are installed in walls or in the space above the ceiling. This will allow for major reworking of the distribution of wiring, ducts, and various items for the building power, lighting, communication, and HVAC systems. The development of details for the construction of both permanent elements and speculatively demountable elements should be made with these provisions in mind.

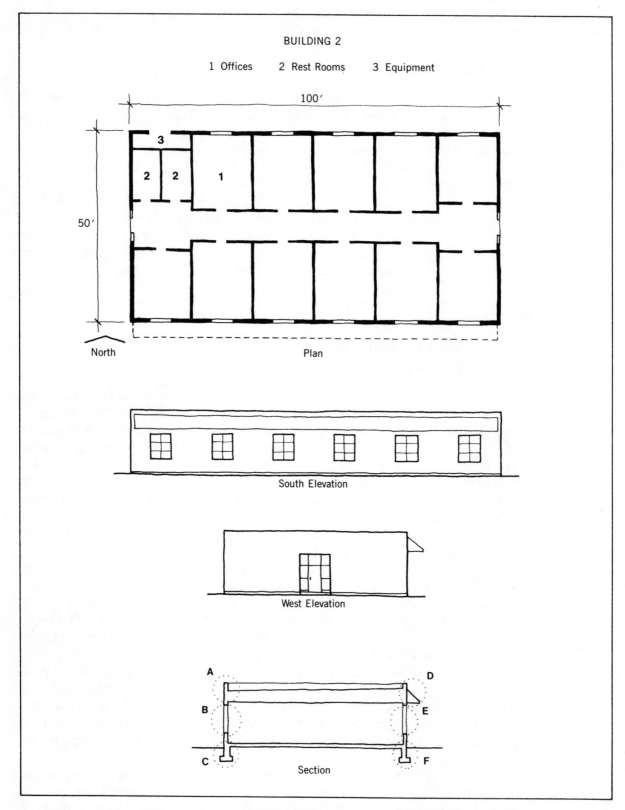

Figure 7.9   Building 2.

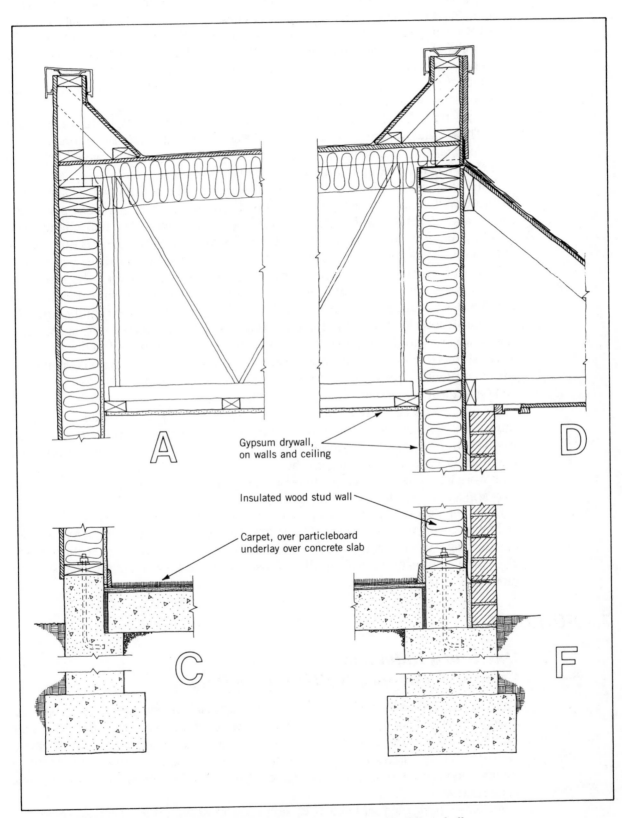

Gypsum drywall,
on walls and ceiling

Insulated wood stud wall

Carpet, over particleboard
underlay over concrete slab

Figure 7.10   Building 2: details of the building shell.

The prefabricated trusses can be fabricated to have a flat bottom chord but a sloping top one to accommodate the roof drainage. For the simplest ceiling construction, the ceiling may be applied directly to the bottom chords of the trusses. With the open webs of the trusses, this still allows for major elements, such as ducts, to run perpendicular to the trusses. This is a case where a deeper than average truss might be used to place the ceiling at a desired level or to provide a maximum of space between truss web diagonals.

An option for the windows, using simple wood-framed elements, is shown in Figure 7.11. These windows are generally not intended to accommodate the intersection of partitions, so the arrangements of any interior partitioning should avoid having the partitions meet the exterior walls except between windows.

Some options for ceiling construction are shown in Figure 7.12. Details are shown for a drywall ceiling at two positions: directly attached to the trusses inside the rooms, and suspended with a separate framing at the hallway. The hall ceiling could also be framed by joists supported by the hallway walls, as the span is quite short. Other forms of ceilings could also be used, supported in any of the three basic ways.

To assure greater sound privacy for the tenants, the wall between the restrooms and the adjacent office might be built with some acoustic enhancement. Some possibilities for this are shown in Figure 7.13. Real sound privacy requires that the total construction be carefully studied during design and inspected during construction to assure that there are no leaks between the spaces intended to be separated. This extends to concern for air ducts, wiring, and other elements that may penetrate the construction. The separating wall should also preferably be extended above the ceiling to meet the underside of the roof deck.

The basic construction shown for this building is quite typical, allowing for regional differences. This is a small building in total floor area, permitting the use of the light wood frame by most building codes. However, the construction may have to develop a one-hour fire rating, depending on the occupancy or on zoning requirements. If the plan were larger, the use of a sprinkler system or the placing of a separating fire wall may still allow the light wood frame to be used.

Depending on the type of occupancy, this building might be developed with no ceiling, exposing the overhead roof structure. This is less likely to occur if any interior partitioning is used, as extending partitions into the truss construction is quite messy. The various problems of dealing with exposed, overhead construction are discussed more fully in other examples.

7.3 BUILDING 3

One-Story Industrial
Concrete/Masonry Walls, Frame Roof

This is a multiple-bayed building, indefinitely extendable in two directions, producing a large covered floor space for an industrial plant, warehouse, or other such usage (see Figure 7.14). Depending on requirements for clear spans, wall heights, height beneath the roof structure, and total floor area, various structural elements may satisfy code requirements, user needs, and the general desire for very economical construction.

As shown here, the roof has a slow-draining, almost flat profile. Drainage of such a roof is a major problem as areas near the center of the large building are some

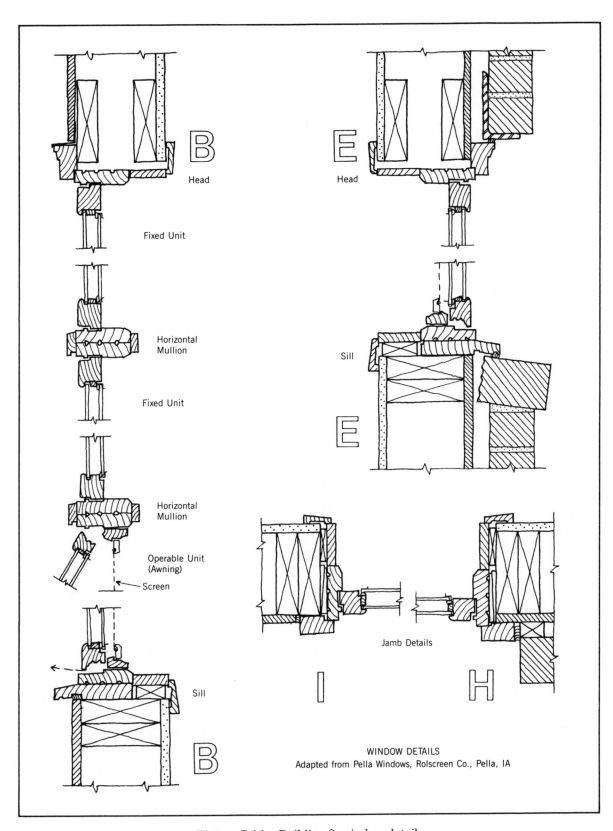

Figure 7.11   Building 2: window details.

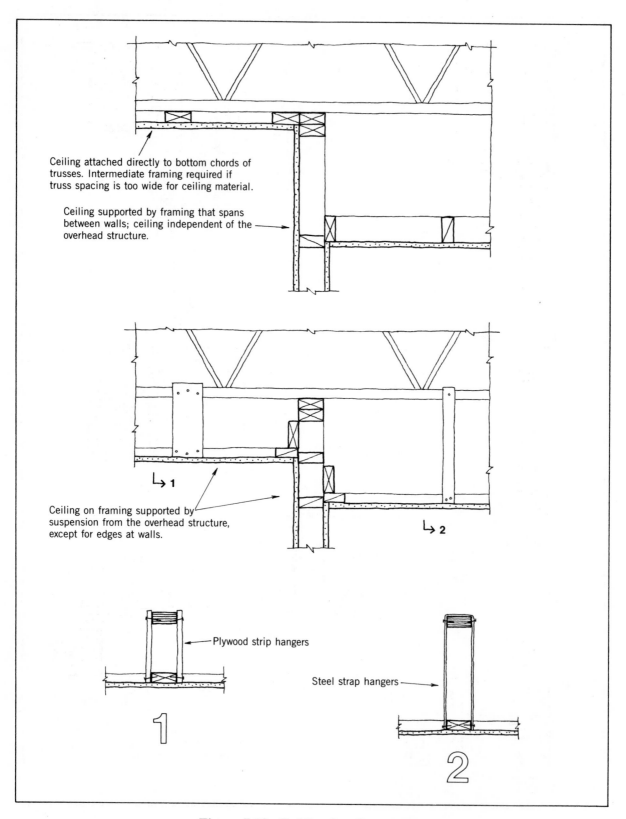

Figure 7.12   Building 2: ceiling details.

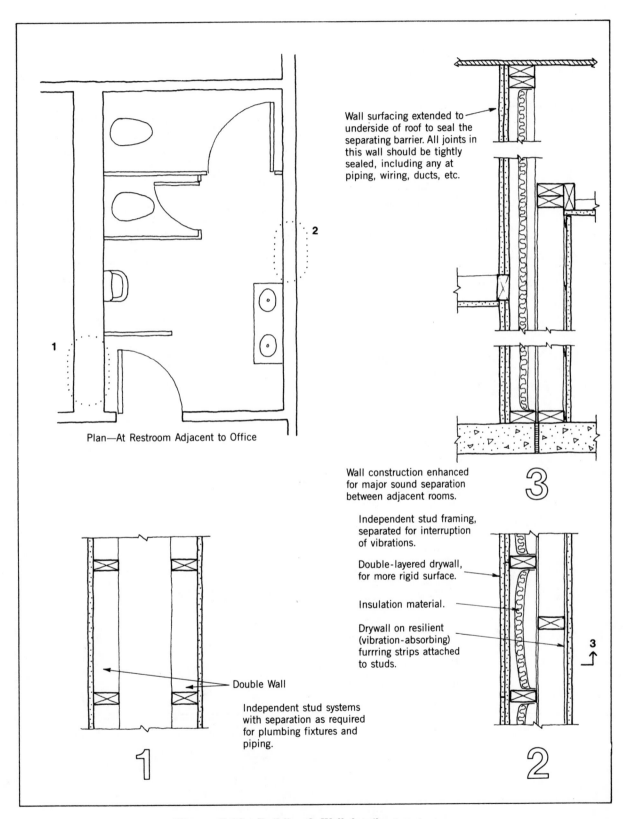

Plan—At Restroom Adjacent to Office

Wall surfacing extended to underside of roof to seal the separating barrier. All joints in this wall should be tightly sealed, including any at piping, wiring, ducts, etc.

Wall construction enhanced for major sound separation between adjacent rooms.

Independent stud framing, separated for interruption of vibrations.

Double-layered drywall, for more rigid surface.

Insulation material.

Drywall on resilient (vibration-absorbing) furrring strips attached to studs.

Double Wall

Independent stud systems with separation as required for plumbing fixtures and piping.

Figure 7.13   Building 2: Wall details at restrooms.

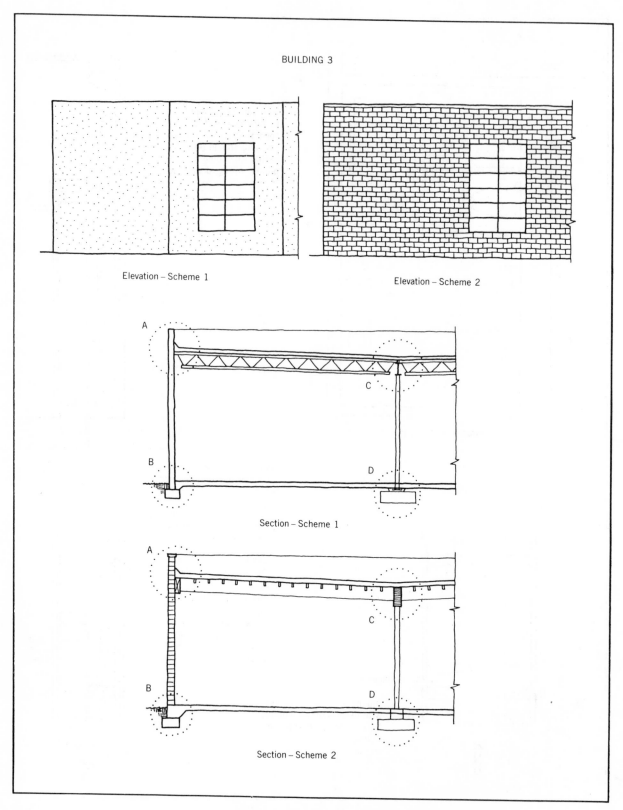

Figure 7.14   Building 3.

distance from a building edge. Edge draining, as shown for building 2, is generally not feasible here, as the total amount of the pitch from the center to an edge would be prohibitive. Some edge draining may be provided, but some roof drains of the type used for interior drainage will be required, most likely discharging into interior vertical leaders located at columns and in turn feeding into an underfloor piped sewer system.

Roof slopes will most likely be achieved here by simply tilting the elements of the roof framing. The nonflat bottom of such a frame is not a concern if no ceiling is provided.

Choice of exterior wall materials will depend on fire requirements, local climate, interior environmental control requirements, need for a particular type of window or for many large windows, concerns for exterior appearance, and other possible concerns. Typically, however, cost will be a major factor, so the most economical option will quite likely be preferred.

A critical factor for the choice of the wall construction will be the height of the wall. These buildings often require considerable interior height; the walls thus become quite tall and must usually span vertically for lateral wind pressure or seismic thrust. The walls may be developed as a framed structure with columns and a curtain wall, or with a major structural form of wall construction.

The two schemes shown in Figure 7.15 use structural walls with potential for use as both bearing and shear walls. In mild climates, the wall construction shown here may be used with no additional materials. In cold climates, however, some enhancement with insulation and moisture barriers will be required, making an insulated sandwich panel system possibly a better choice.

Options for the roof structure are considerable in number, and the most economical system will probably be chosen. The two systems shown here are common and can be used for a range of spans. If appearance of the exposed structure is of some concern, it is desirable to have the necessary finishing of the trusses done in the fabricating shop, as painting them at the site is quite laborious.

Floor slabs are usually left uncovered, although surfaces may be treated in some manner to produce a better than usual hard, smooth face for wear resistance. Any floor treatment is possible, but rough use and economics usually dictate the use of the exposed concrete surface. Various surfaces can be developed by working the concrete as it hardens or applying materials after the concrete has cured.

While the general interior of such a building may be as shown in Figure 7.15, there is usually some need for partitioned spaces within the general enclosure—for toilets, offices, equipment, and so on. Figure 7.16 presents some details for the creation of such spaces that include the development of partitions, a suspended ceiling, and some enhancement of the inside surface of the exterior wall. Floor treatments may also be used in such a space, although the change in finished floor surface level will present a minor problem at doorways between the two types of spaces, which probably favors use of relatively thin flooring materials.

7.4 BUILDING 4

Two-Story Motel
Masonry and Precast Concrete

As shown in Figure 7.17, this building has a common form of plan for motels, with rooms off a central corridor. The same basic structure might also be used for a dormitory or small apartment building.

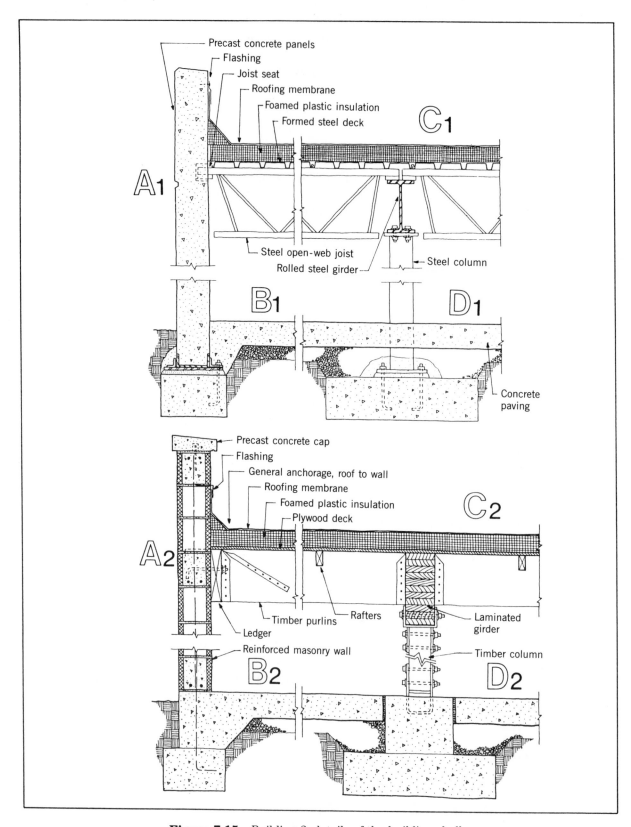

Figure 7.15   Building 3: details of the building shell.

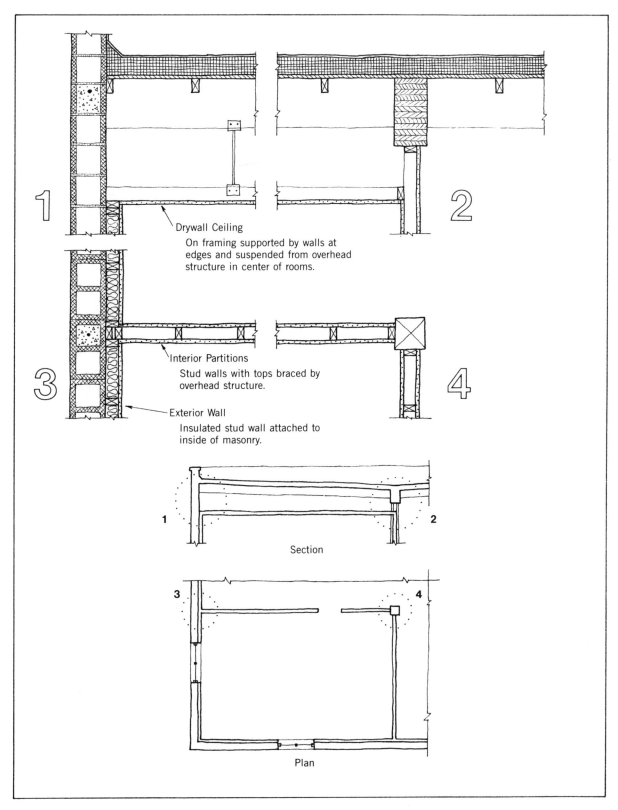

Figure 7.16 Building 3: details for development of finished interiors.

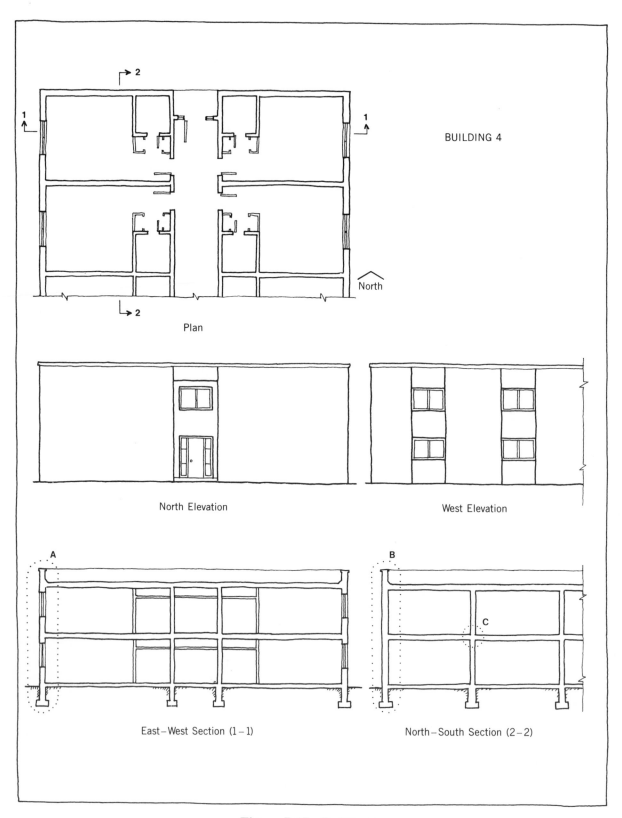

Figure 7.17 Building 4.

The basic structure, as shown in Figure 7.18, consists of structural masonry walls and precast concrete roof and floor units. Although this is a widely used construction, there are many possibilities for variations, both in the general structure and in finish materials and details. Higher quality (supporting higher room charges) is usually achieved with better finish materials, although some quality may be achieved by enhancements of the basic construction as well.

Choices for the masonry wall structure depend largely on regional consideratons, involving both climate and type of critical lateral force effects. The exterior wall construction shown here utilizes reinforced concrete block (CMUs) as the structure with an exterior brick veneer. This is a system favored in areas with severe windstorms or high seismic risk, and the general form of the structure would change in other regions where seismic response is not critical.

In cold climates the masonry walls need considerable enhancement to improve the character of thermal flow. The uninsulated walls shown here would be possible only in a very mild climate. Insulation can be developed in a variety of forms, including the following:

Foam plastic units adhered to the inside surface of the wall with gypsum drywall adhered to the insulation units

Glass fiber batt insulation in a void space created by furring out on the inside of the masonry, with gypsum drywall fastened to the furring strips

Foam plastic insulation in the cavity space between the CMUs and the brick veneer

Foam plastic units adhered to the outside face of the masonry (without the brick veneer) with a priority vinyl stuccolike finish with fiberglass mesh reinforcement

The development of interior wall finishes must first relate to any considerations for the exterior walls, including those just described. The walls at the windows are developed as light wood frame, with all of the usual options for surfacing on the inside of the frame.

Development of interior walls is shown in Figure 7.19. The walls between motel rooms are structural CMU construction, providing support for the roof and floor units. These are relatively heavy walls and may well be left exposed, possibly receiving only some paint. However, for a higher-quality finish, they may receive some surfacing with paneling or plaster. Their sound privacy quality may also be enhanced by using a furred-out surface on one side, with an additional degree achieved by using resilient attachment of the surfacing to the furring.

The nonstructural partitions for the halls, bathrooms, and closets would most likely be achieved with simple light wood framing. A dropped ceiling is typically used in these areas to allow for encasing of air ducts, wiring, and piping. This form of construction is shown in the details in Figure 7.19.

7.5 BUILDING 5

Church Auditorium/Chapel
Exposed Timber Structure

This is a single-space, medium-span building with essentially no interior structure. The options possible for this building in terms of type of structure and architectural style are virtually endless. Shown here is a popular form of construction, with gabled

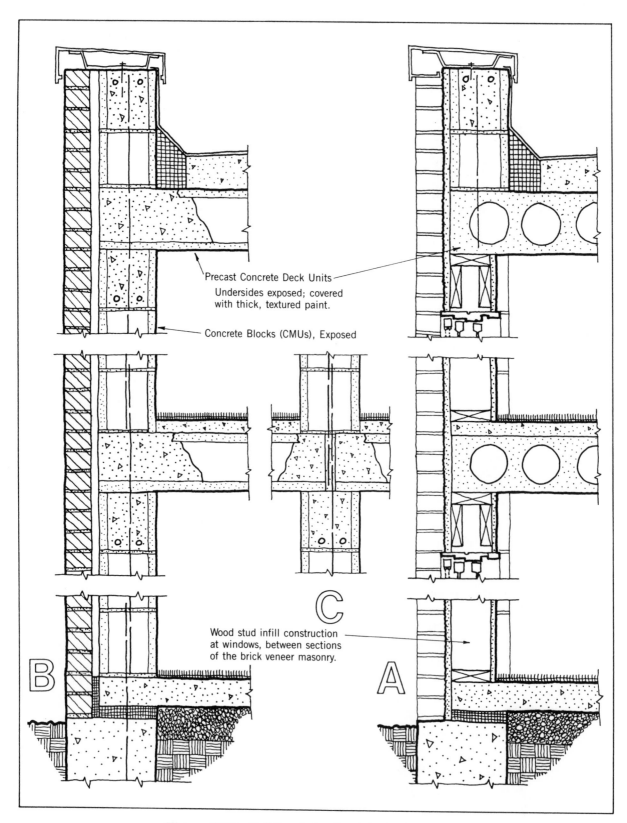

Precast Concrete Deck Units
Undersides exposed; covered
with thick, textured paint.

Concrete Blocks (CMUs), Exposed

Wood stud infill construction
at windows, between sections
of the brick veneer masonry.

C

B

A

Figure 7.18 Building 4: details of the building shell.

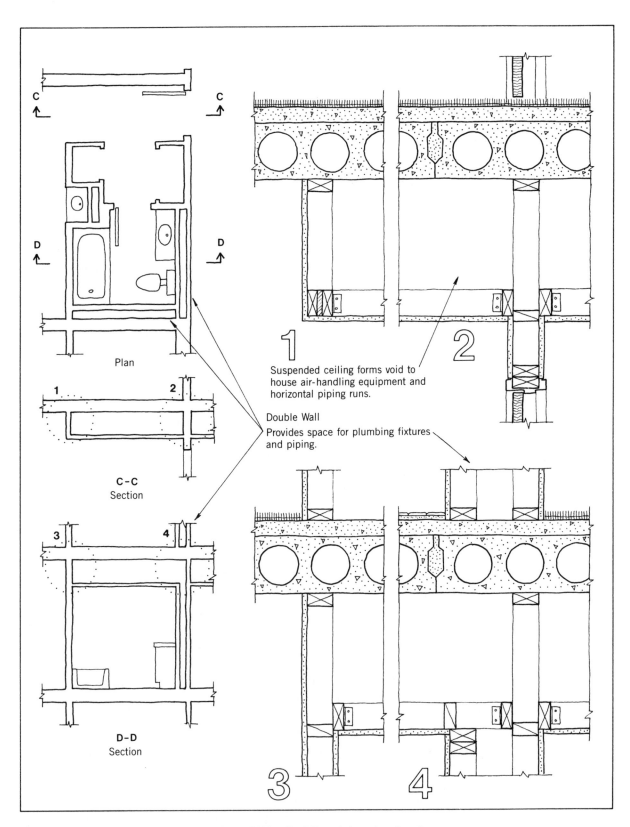

Figure 7.19   Building 4: interior details.

bents of glued laminated wood and an exposed wood beam-and-plank deck roof. Infill walls between the gabled bents are achieved with light wood frame construction.

The general form of the building and the vaulted interior are shown in Figure 7.20. The front podium/altar area is slightly raised as a stepped platform. The other end of the church may well contain some rooms and a balcony with an entry area and stairs, but we will deal only with the portion of the building shown here.

The basic construction of the building shell is shown in the details in Figure 7.21. The building interior is comprised primarily of the insides of the roof and exterior walls and the top of the concrete pavement. Further development of the interior consists mostly of choices for the finishes on these elements.

The details of the side windows are shown in Figure 7.22, indicating the use of wood-framed sash and wood trim carrying out the use of exposed wood materials on the inside of the building. This could be carried another degree by using wood paneling on the wall surfaces and wood flooring, but it might smack of a lumberyard display and be a bit of overkill for natural materials.

Development of the podium/altar area will depend very much on the form of services by the church group using the building. It may also need to respond to other uses for the building if it doubles as a general-purpose auditorium for meetings, concerts, plays, and so on. The steps and platform can be achieved in various ways, two common ones being those shown in Figure 7.23.

The wood platform is simply built up on top of the concrete paving slab. This is undoubtedly the cheap way to go, although construction can be made as sturdy as desired. An advantage for this construction is the possibility of future alteration with relative ease should the users of the building develop different needs. The wood construction should be well anchored and joists should be extra stiff to eliminate bounciness. The deck should also be extra stiff (not pushing limits of recommended thicknesses) and should be held down with screws and glue to assure no squeaking.

The alternative construction shows a stepped concrete structure. This is a much sturdier, bounceproof form of construction. One reason for choosing it might be the choice of tile floor surfacing, as shown in the details. This provides a much better support for the tiles, although either finish could actually be placed on either of the structures shown here.

This is an example of a situation where the basic construction and actual structural elements are hung out for display, and the designer of the interior must anticipate the various details and finishes of the materials that can normally be expected. Tolerance for the normal range of acceptable dimensional inaccuracies must be developed in the construction that abuts and infills the basic structure. For example, at the top of the wall in detail B and the ends of the walls in detail D, some form of molding or reveal should be used to achieve the joint between the two planes of different materials, so as not to have just a blind collision.

Jointing of the timber frame, grade specifications for the timber elements, and side finish for the glued laminated bents should be done with good information on local products and fabricating practices. Joints may be boldly developed with exposed heavy steel elements, or more cleanly developed with some concealed fastening devices. The general impact of these details on the whole interior design should be carefully studied.

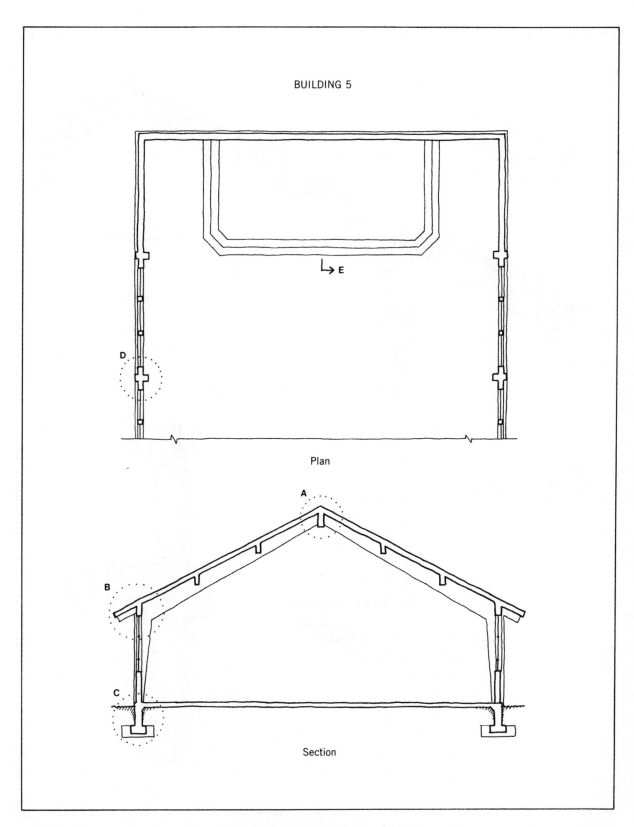

BUILDING 5

Plan

Section

Figure 7.20 Building 5.

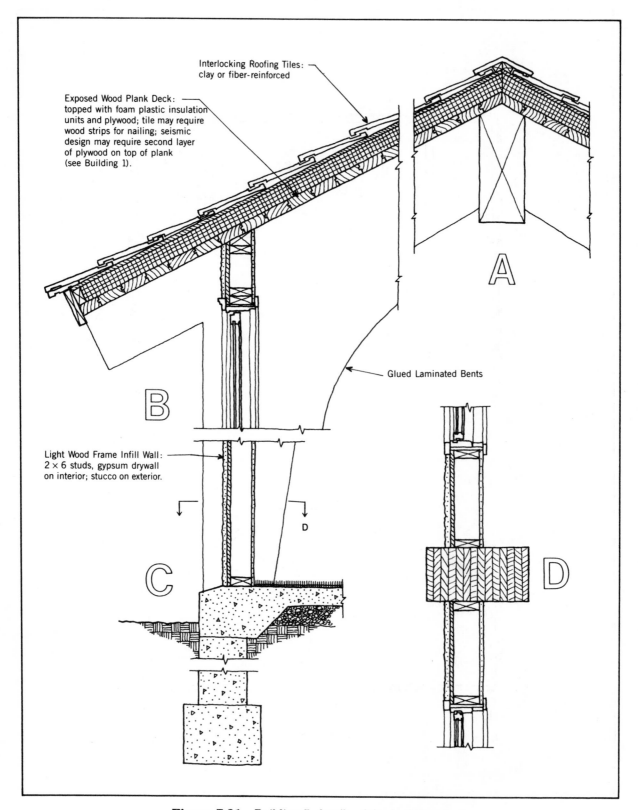

Figure 7.21   Building 5: details of the building shell.

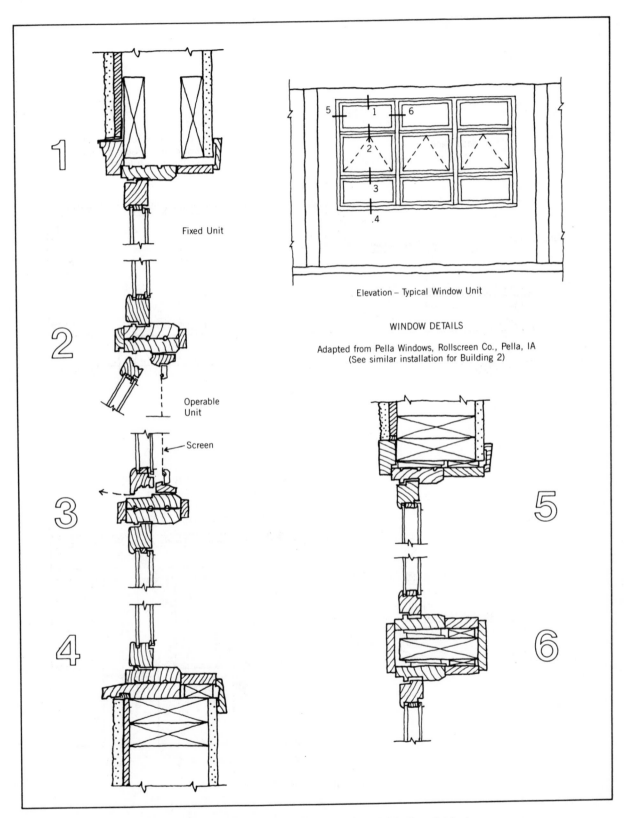

Fixed Unit

Operable
Unit

Screen

Elevation – Typical Window Unit

WINDOW DETAILS

Adapted from Pella Windows, Rollscreen Co., Pella, IA
(See similar installation for Building 2)

Figure 7.22   Building 5: details of the infill wall and windows.

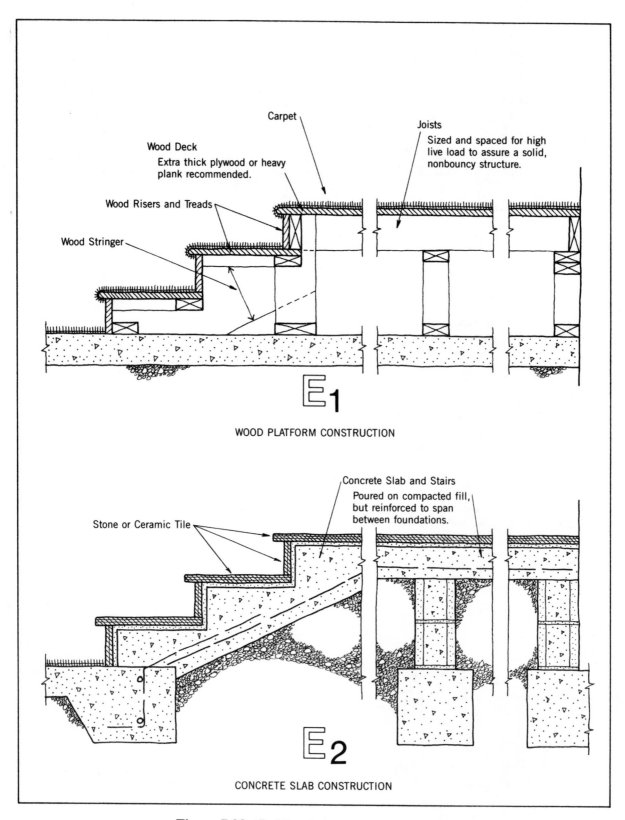

Figure 7.23   Building 5: details for the podium.

7.6 BUILDING 6

Three-Story Office Building
Steel Frame, Metal Curtain Wall

This is a modest-size building, generally qualified as being low rise (see Figure 7.24). In this category there is a considerable range of choice for the construction, although in a particular place, at a particular time, a few popular forms of construction tend to dominate the field. Currently available products, market competition, current code requirements, popular architectural styles, and the general preferences of builders, craft people, developers, and designers usually combine to favor a few basic forms of general construction.

Shown here is a steel frame structure with W sections used for columns, girders, and beams, and a formed sheet steel deck for the floors and roof. Many variations are possible with this basic system, mostly dealing with choices for the deck system and the members that directly support the deck.

The wall system here is a curtain wall developed as an infill steel stud system. Windows consist of strips between columns, framed into rough openings in the stud wall structure, not unlike the basic method used in buildings 1, 2, 4, and 5. The basic form of the curtain wall system is shown in Figure 7.25.

Details of the basic enclosure system are shown in Figure 7.26. For development of the interior, some relationships with the exterior wall and structure are as follows:

1. Interior finish of the exterior wall is part of the curtain wall development, but also an interior surface that must be coordinated with other interior construction. Shown here is a simple gypsum drywall, but many other options are possible, including application of various finishes to the drywall surface.
2. Positioning of the column with relation to the exterior wall and spandrel establishes the nature of the column's presence in the building interior. In this case the wall is thickened to enclose the column, so its only interior presence is in the form of the width of space it occupies in the wall. For comparison, see the situation with the exterior columns in building 7, as shown in Figures 7.30 and 7.31.
3. The locations of vertical window mullions represent potential locations for interior partition walls that intersect the exterior wall. This calls for some design coordination with the planning of interior walls and possibly with any modular ceiling, lighting, and so on.
4. Other details of the windows will also relate to development of interior spaces, such as sill height, head height, presence of horizontal mullions, space for installation of drapes or shades, form of operation of operable sash, and so on.

It is common to use some basic modular planning in this type of building. Spacing of the building columns constitutes one large unit in this regard, but ordinarily leaves a wide range for smaller divisions. Modular coordination may also be extended to development of ceiling construction, lighting, ceiling HVAC elements, fire sprinklers, and the systems for access to electric power, phones, and other signal wiring systems.

There is no single magic number for an interior modular planning system, and all dimensions between 3 and 5 ft have been used for partitions and window mullions. In the United States, common modular units are 4 ft, 12 in., and 4 in. Most priority ceiling systems and corresponding lighting systems use the 1-ft/2-ft/4-ft system for basic

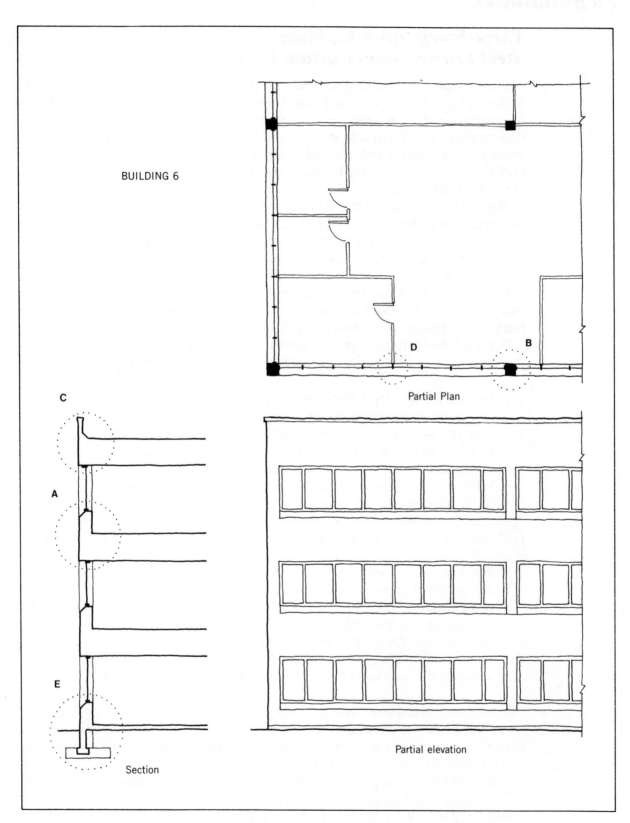

BUILDING 6

Partial Plan

Section

Partial elevation

Figure 7.24   Building 6.

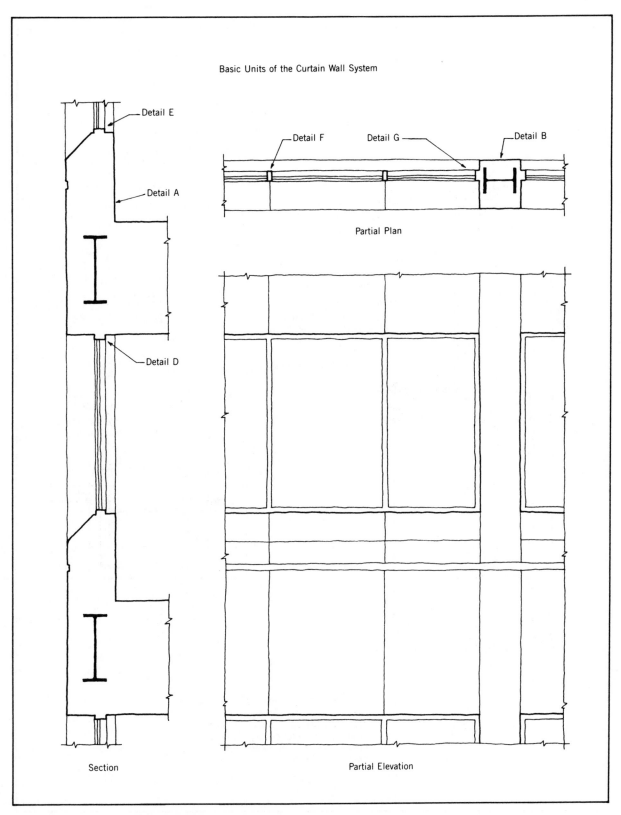

Figure 7.25   Building 6: basic units of the curtain wall system.

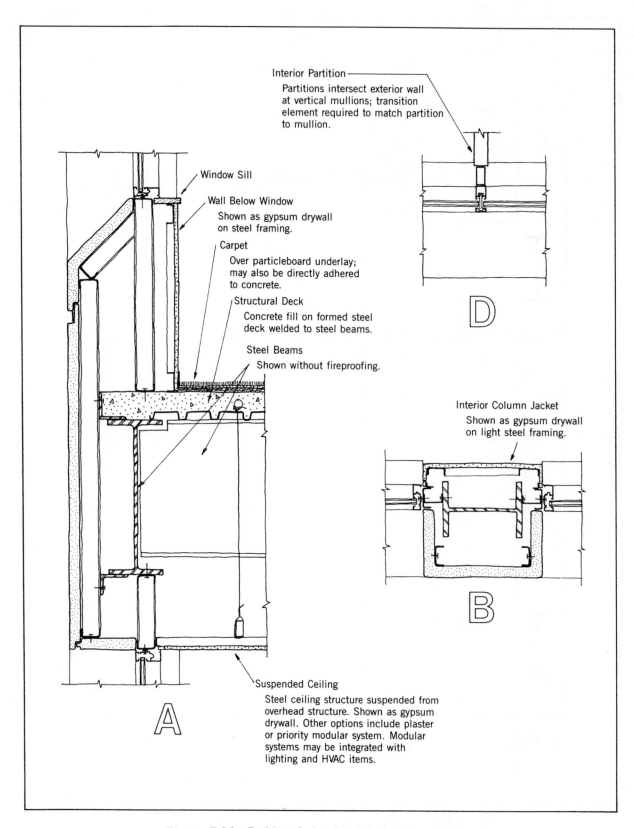

Interior Partition

Partitions intersect exterior wall at vertical mullions; transition element required to match partition to mullion.

Window Sill

Wall Below Window

Shown as gypsum drywall on steel framing.

Carpet

Over particleboard underlay; may also be directly adhered to concrete.

Structural Deck

Concrete fill on formed steel deck welded to steel beams.

Steel Beams

Shown without fireproofing.

Interior Column Jacket

Shown as gypsum drywall on light steel framing.

Suspended Ceiling

Steel ceiling structure suspended from overhead structure. Shown as gypsum drywall. Other options include plaster or priority modular system. Modular systems may be integrated with lighting and HVAC items.

D

B

A

Figure 7.26   Building 6: details of the building shell.

elements. However, 4-ft elements can be spaced on 5-ft centers or otherwise utilized in other planning modules. Selection of a particular manufacturer's system for interior development may establish some criteria for this planning.

For buildings built as investment properties, with speculative occupancies that vary over the life of the building, it is usually desirable to accommodate future redevelopment of the building interior with some ease. For the basic construction, this means a design with as few permanent structural elements as possible. At a bare minimum, what is usually required is the construction of the major structure (columns, floors, roof), exterior enclosing walls, and interior walls around stairs, elevators, restrooms, equipment rooms, and risers for building services. Everything else preferably should be nonstructural and easily demountable, if possible.

The space between the underside of suspended ceilings and the top of floor or roof structures must typically contain many elements besides those of the basic construction. This represents a situation requiring major coordination for the integration of the space needs for the elements of the structural, HVAC, electrical, communication, lighting, and fire-fighting systems. A major design decision that must often be made very early in the design process is that of the overall dimension of the space required for this collection of elements. Depth permitted for the spanning structure and the general level-to-level vertical building height will be established—and not easy to change later, if the detailed design of any of the enclosed systems indicates a need for more space.

Generous provision of the space for building elements will make the work of the designers of the various subsystems easier, but the overall effects on the building design must be considered. Extra height for the exterior walls and permanent interior walls, and for stairs, elevators, and service risers all result in additional cost, making tight control of the level-to-level distance very important.

As previously mentioned, a detail consideration in planning the interior is the manner in which interior partitioning meets the exterior walls. Partitions at right angles to the exterior must join to a column, a solid portion of wall, or—in some manner—to a vertical framing member in the window system. The construction of the partitions and of the columns, windows, and other exterior wall elements must be studied in this regard. Some of the details in Figure 7.26 reflect the need for this concern.

Obviously, many other options are possible for the general form, the materials, and the details of the construction of the exterior walls and the elements of the building structure that occur at the building edge. Figure 7.27 shows some variations on the building edge condition and the ramifications with regard to development of interior partitioning and ceilings.

Details for the windows, consisting of standard curtain wall units, are shown in Fig. 7.28. These are units made for horizontal strip window development and not for a general curtain wall system such as that shown for building 7.

Suspended ceilings, as shown in Fig. 7.29, are usually developed with a basic structure that permits some future changes with regard to redevelopment of interior designs for new occupants. Suspension methods must be related to the overhead construction and to the specific finished form of the ceiling in any case, but generalized techniques are most effective. The details in Figure 7.29 show the use of a basic suspension system for the support of a simple drywall ceiling.

Interior partitions have many possible variations, all of which must respond to code requirements for fire separation and to various occupant needs for privacy, security,

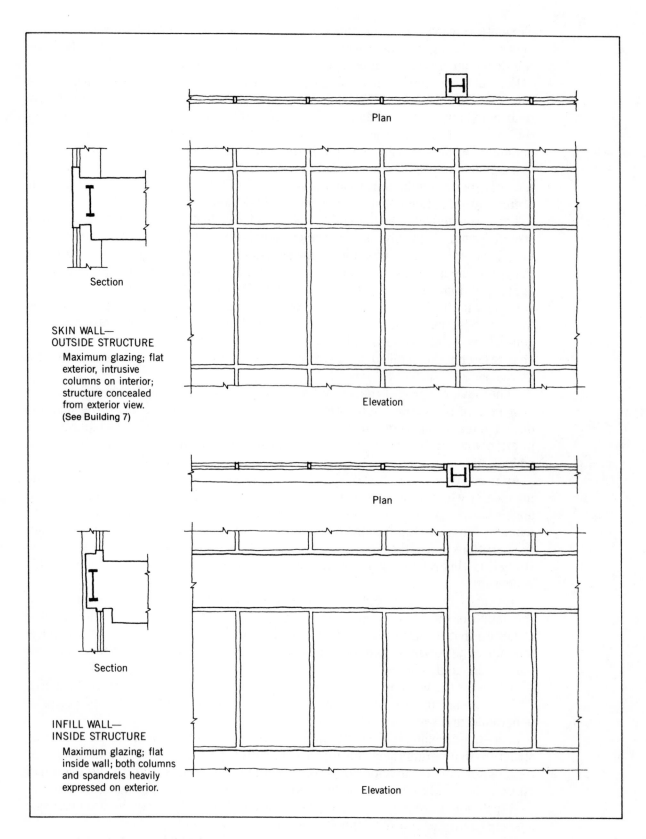

Plan

Section

SKIN WALL—
OUTSIDE STRUCTURE

Maximum glazing; flat
exterior, intrusive
columns on interior;
structure concealed
from exterior view.
(See Building 7)

Elevation

Plan

Section

INFILL WALL—
INSIDE STRUCTURE

Maximum glazing; flat
inside wall; both columns
and spandrels heavily
expressed on exterior.

Elevation

Figure 7.27   Building 6: alternative details for the development of the building exterior walls.

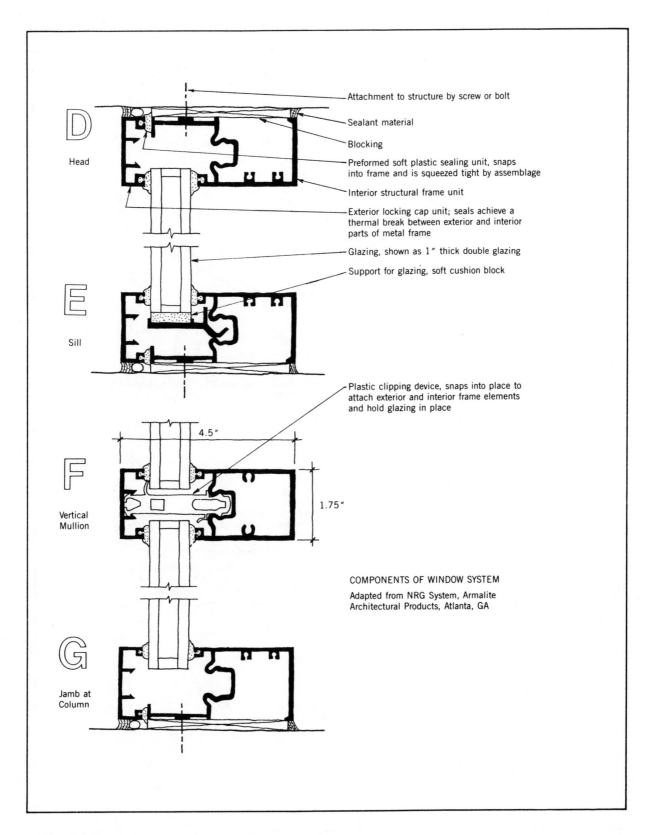

Figure 7.28   Building 6: typical elements of the curtain wall system. Adapted from the NRG System, produced by Armalite Architectural Products, Atlanta, Georgia.

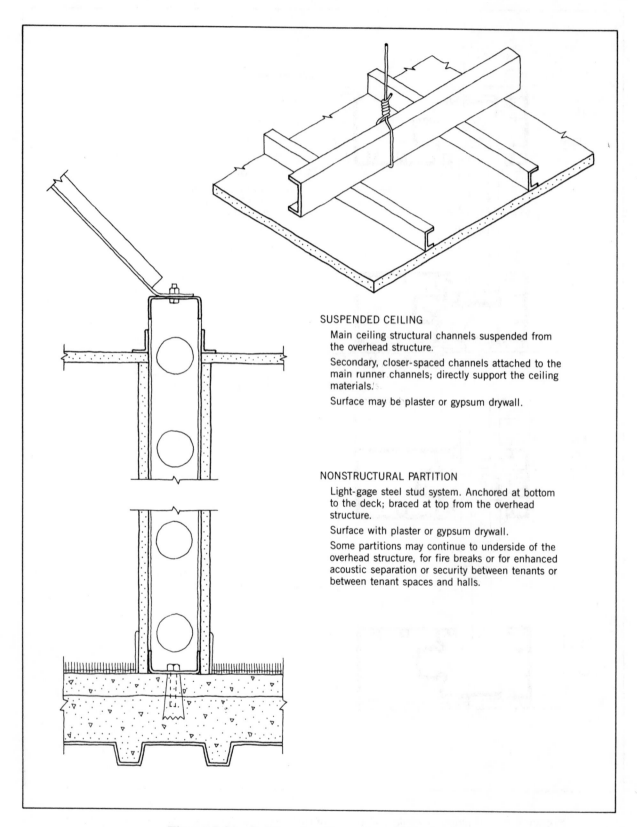

SUSPENDED CEILING

Main ceiling structural channels suspended from the overhead structure.

Secondary, closer-spaced channels attached to the main runner channels; directly support the ceiling materials.

Surface may be plaster or gypsum drywall.

NONSTRUCTURAL PARTITION

Light-gage steel stud system. Anchored at bottom to the deck; braced at top from the overhead structure.

Surface with plaster or gypsum drywall.

Some partitions may continue to underside of the overhead structure, for fire breaks or for enhanced acoustic separation or security between tenants or between tenant spaces and halls.

Figure 7.29   Building 6: ceiling and partition details.

and use of walls for display, storage, and so on. Walls must also support doors, and the construction should be developed with this in mind.

Not to dismiss it as trivial, but the generally easiest things to change are the finishes of interior surfaces. Major changes in interior appearance can be effected by alteration of floor, wall, and ceiling surfaces—usually with no major effects on the general construction. Some accommodation of this may be heightened by choices for the basic construction, but the typical situation allows for considerable refinishing. Our concern here is basically for construction, not for interior decoration; so we will not deal further with these matters.

The general development of the construction for this building should be compared with that for building 7. There are major differences—in exterior appearance, planning, and various design concerns.

7.7 BUILDING 7

Ten-Story Office Building
Concrete Frame, Metal Curtain Wall

This building (see Figure 7.30) is not exactly high-rise, but it is enough taller than building 6 to result in some narrowing of choices for the structure and some different planning concerns. It is unlikely that anything other than a steel or concrete frame would be used for this building.

The type of curtain wall used here (see Figure 7.31) is one that forms a continuous skin outside the building structure. The columns at the building edge are thus almost freestanding inside the wall, a plan intrusion in the occupied space that some designers or owners might find objectionable. For comparison, see the plan and wall details for building 6, in which the columns are enclosed by the thick exterior wall at the floor level.

Formation of the curtain wall is achieved using the "stick" system, in which vertical structural mullions are developed like a stud system to provide basic support for the rest of the wall. Windows continue in horizontal rows past the columns, so that the building exterior surface does not reveal the pattern of the column spacing, as it does in building 6.

Spandrel covers are placed in the same plane with the windows and have glass on the exterior surface. A continuous all-glass wall is thus produced, and if dark or reflective glazing is used, the difference between glazing and spandrel panels may not be apparent in daylight. The spandrels thus also become obscured, and a generally undelineated structural surface results, with neither vertical nor horizontal structural references visually apparent.

Details of the general construction shown in Figure 7.32 indicate the form of the curtain wall and the possibilities for development of interior partitioning that abuts the exterior wall. As with building 6, there are two general cases: partitions that meet a column and partitions that meet a vertical mullion in the curtain wall. The difference here is that there is really no solid portion of wall as far as the interior partitions are concerned, as the vertical mullions are essentially continuous from floor to ceiling.

The plans, sections, and details reveal the use of the sitecast concrete structure. The horizontal spanning system consists of a solid two-way-spanning slab, with beams used only at the edges and at large openings on the interior. One advantage of this system is the possibility for reduction of the space enclosed by the ceiling below

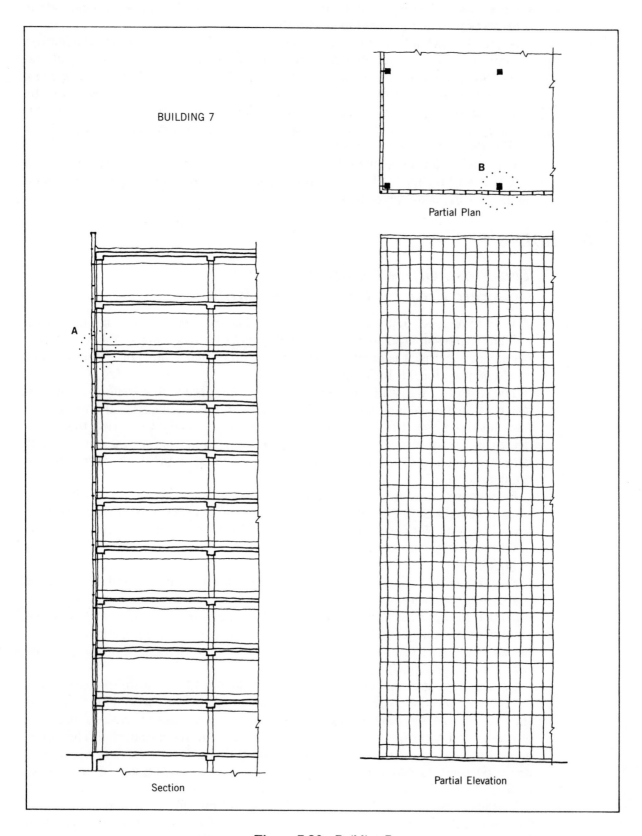

BUILDING 7

Partial Plan

Section

Partial Elevation

Figure 7.30   Building 7.

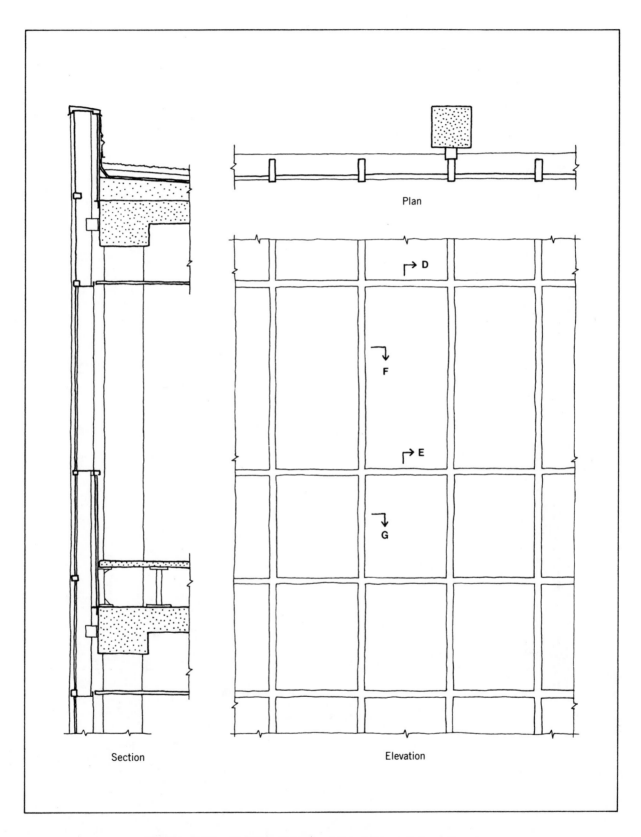

Plan

Section

Elevation

Figure 7.31   Building 7: basic units of the curtain wall system.

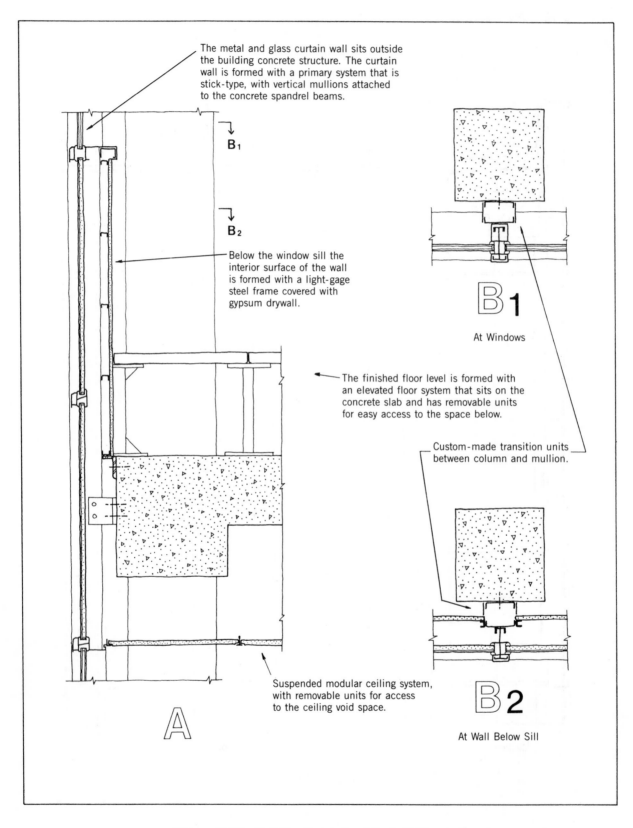

The metal and glass curtain wall sits outside the building concrete structure. The curtain wall is formed with a primary system that is stick-type, with vertical mullions attached to the concrete spandrel beams.

B_1

B_2

Below the window sill the interior surface of the wall is formed with a light-gage steel frame covered with gypsum drywall.

The finished floor level is formed with an elevated floor system that sits on the concrete slab and has removable units for easy access to the space below.

Custom-made transition units between column and mullion.

Suspended modular ceiling system, with removable units for access to the ceiling void space.

A

B 1

At Windows

B 2

At Wall Below Sill

Figure 7.32   Building 7: details of the building shell.

and the structure above, as beams do not regularly project into this space as with ordinary slab-and-beam framing. Thus the space available for ducts and other service elements is that bounded by the top of the ceiling system below and the underside of the slab above.

Development of the suspended ceiling is shown in Figure 7.33. With this structure, it is also conceivable that the suspended ceiling might be eliminated, as the thick concrete slab does not require other fire protection. Thus the underside of the structure could be exposed and the necessary overhead services hung out for display, a high-tech style of design popular with some designers.

For this building, the drawings also indicate the use of an elevated floor system, consisting of a platform working floor surface developed by a system of modular panels supported by closely spaced, short posts (see Figure 7.34). This creates an interstitial space that can be used to incorporate various building service elements, most commonly the wiring for power, phones, computer networks, and other signaling systems. For occupancies in which major use of communication and computing systems is made—and frequently modified—this is a highly useful construction.

The heavy concrete slab is also one of the best systems for the accommodation of the concentrated loads from the supports for the elevated floor system. Lighter decks may have difficulty with these concentrated, punching, shear types of loads. However, office floor loadings from building codes typically require some concentrated live load effect, so this may be less critical for office buildings.

Partitions may be developed in various ways with the elevated floor system. Some priority systems may match up the two—floor and partitions—with special details. Permanent walls and other full-height partitions may be simply built on top of the concrete slab (see Figures 7.33 and 7.34). Partial-height partitions, used for office landscape developments, can be supported on top of the elevated floor, although the specific recommendations of the manufacturer of the elevated floor system should be followed.

7.8 BUILDING 8

Sports Arena, Longspan Roof
Steel Two-Way Truss System

This is a medium-size sports arena, possibly big enough for a swim stadium or a basketball court (see Figure 7.35). Options for the structure here are strongly related to the desired building form. Functional planning requirements derive from the specific activities to be housed and from the seating, internal traffic, overhead clearance, and exit and entrance arrangements.

In spite of all of the requirements, there is usually some room for consideration of a range of alternatives for the general building plan and overall building form. Choice of the system shown here relates to a commitment to a square plan and a flat roof profile. Other choices for the structure may permit more flexibility in the building form or also limit it. Selection of a dome, for example, would require a round plan, or at least one that could be developed under the hemispherical enclosure. Need for an oblong plan would limit the possibility for a two-way spanning structure.

The structure shown here uses a two-way spanning steel truss system (often called a "space frame," although the name is quite ambiguous). Basic planning with this type of structure requires the use of a module that relates to the nodal points (joints) of the truss system. Locations of supports must relate to the nodal points, and any major

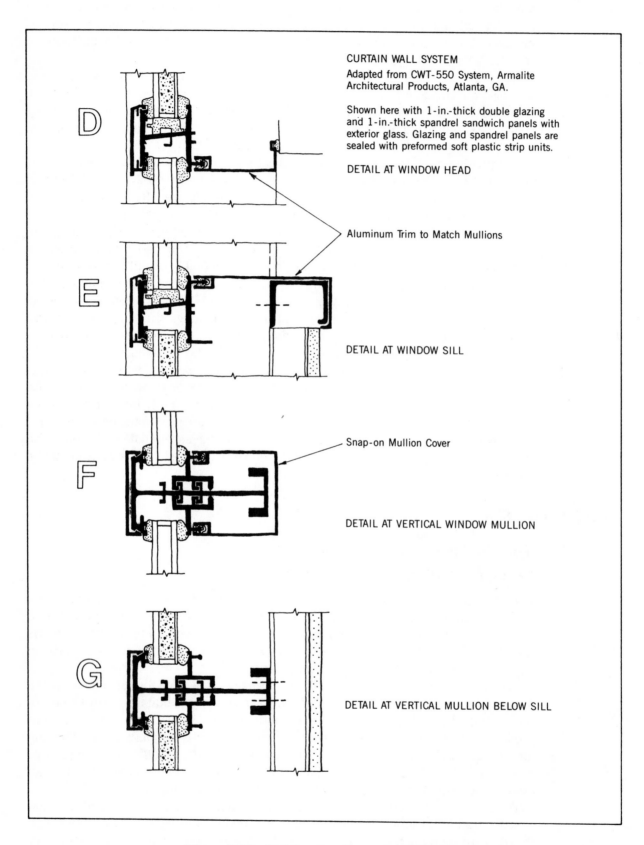

CURTAIN WALL SYSTEM

Adapted from CWT-550 System, Armalite
Architectural Products, Atlanta, GA.

Shown here with 1-in.-thick double glazing
and 1-in.-thick spandrel sandwich panels with
exterior glass. Glazing and spandrel panels are
sealed with preformed soft plastic strip units.

DETAIL AT WINDOW HEAD

Aluminum Trim to Match Mullions

DETAIL AT WINDOW SILL

Snap-on Mullion Cover

DETAIL AT VERTICAL WINDOW MULLION

DETAIL AT VERTICAL MULLION BELOW SILL

Figure 7.33   Building 7: ceiling and partition details.

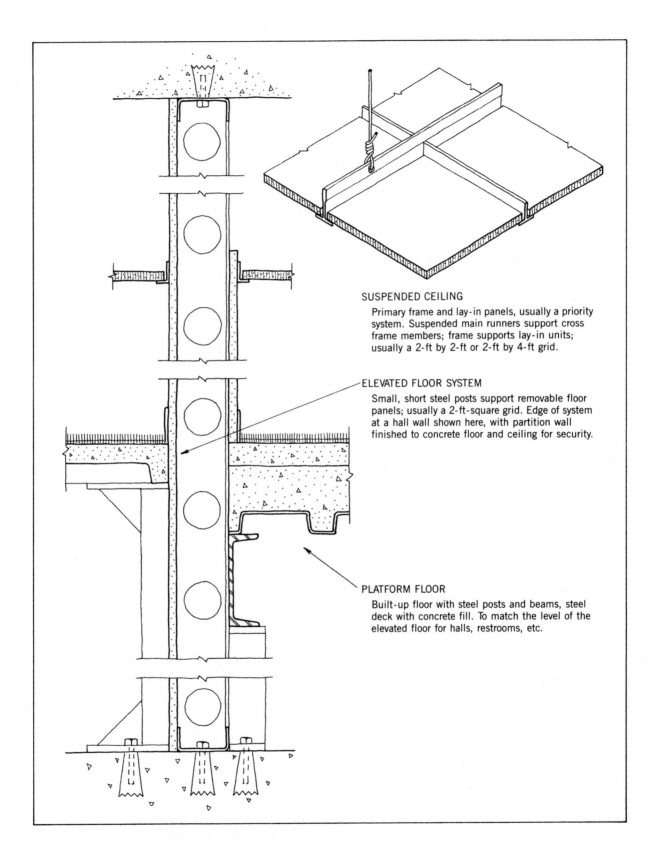

SUSPENDED CEILING
Primary frame and lay-in panels, usually a priority system. Suspended main runners support cross frame members; frame supports lay-in units; usually a 2-ft by 2-ft or 2-ft by 4-ft grid.

ELEVATED FLOOR SYSTEM
Small, short steel posts support removable floor panels; usually a 2-ft-square grid. Edge of system at a hall wall shown here, with partition wall finished to concrete floor and ceiling for security.

PLATFORM FLOOR
Built-up floor with steel posts and beams, steel deck with concrete fill. To match the level of the elevated floor for halls, restrooms, etc.

Figure 7.34   Building 7: details of the elevated floor.

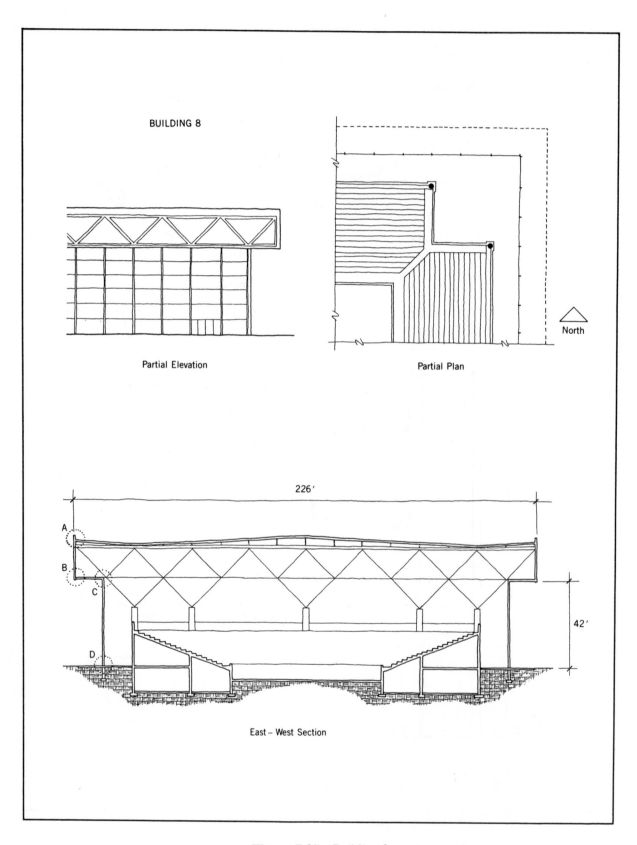

Figure 7.35 Building 8.

concentrated loads should be applied at nodal points. The use of the nodal-point module may be extended to general planning of the building, as has been done in this example.

There is nothing particularly unique about the construction shown here, but it cannot really be called common or standard. Even in unique buildings, however, it is common to use as many standard products as possible. Thus the roof structure on top of the truss and the curtain wall system for the exterior walls use off-the-shelf products (see Figure 7.36).

It is possible that the truss system shown here could be produced with various priority systems, and it is generally advisable to pursue these when beginning the design of such a structure. Use of such a system will eliminate a great deal of design effort and provide some greater assurance of acceptable performance of the structure. The details shown here indicate an ordinary truss system using rolled steel shapes and steel plate gussets; a possibility among various options. For the exposed structure another popular form is that produced with steel pipes—in the manner of a giant bicycle frame.

In addition to the longspan structure in this case, there is also the major problem in developing the 42-ft-high curtain wall. Braced laterally at only the top and bottom, this is a 42-ft span structure sustaining wind pressure as a major load. With even modest wind conditions, producing pressures in the 20-psf range, this requires some major vertical structure to span the 42 ft.

In this example, the fascia of the roof trusses, the soffit of the overhang, and the curtain wall surface are all developed with products ordinarily available for curtain wall construction. The vertical span of the tall wall is developed by a series of custom-designed trusses that establish a major spacing module for the vertical mullions, analogous to the columns in building 7. A horizontal structure is developed to span the distance of 14 ft between the vertical trusses—steel structural tubing in this case. The steel tubes then provide both vertical support and horizontal bracing for the 3.5-ft-wide by 7-ft-high window units.

The inside of the roof and curtain wall shell and the exposed steel truss system strongly define the building interior. The truss members define a pattern that is orderly and pervading. There will, however, most likely be additional items overhead, within, and possibly beneath the trusses. These may include:

Drains and piping for the roof drainage system

Ducts and registers for the building HVAC system

Wiring and fixtures for a general lighting system

Elements required for a general audio system

Signs, scoreboards, etc.

Catwalks for access to the various overhead equipment

To preserve some design order, as much of the overhead equipment as is possible should be developed within the truss modular system.

The building shell, of course, exists only to house the building activities. In this case, an essentially independent construction is developed under the large overhead enclosure to form the seating and other required facilities. The support columns for the truss are incorporated into this construction, but the separation is otherwise

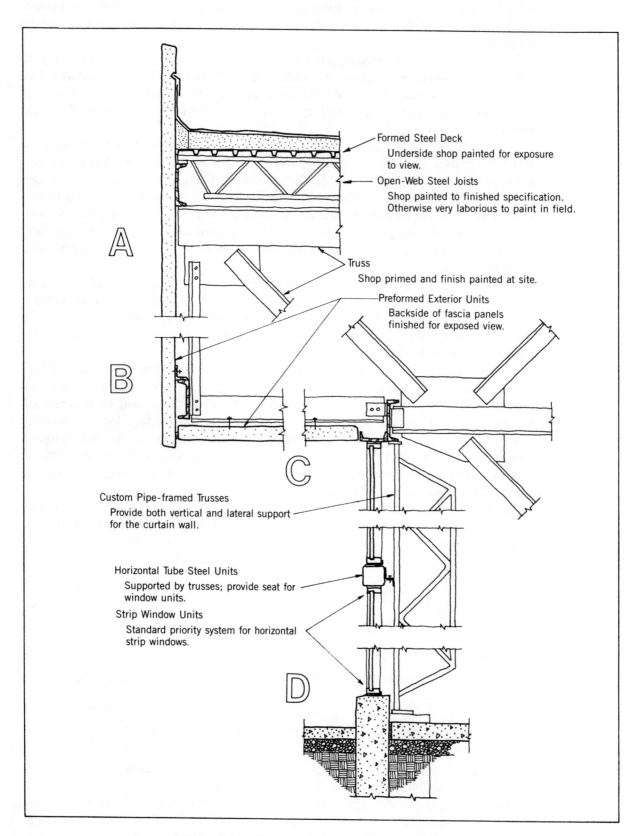

Formed Steel Deck
Underside shop painted for exposure to view.

Open-Web Steel Joists
Shop painted to finished specification. Otherwise very laborious to paint in field.

Truss
Shop primed and finish painted at site.

Preformed Exterior Units
Backside of fascia panels finished for exposed view.

Custom Pipe-framed Trusses
Provide both vertical and lateral support for the curtain wall.

Horizontal Tube Steel Units
Supported by trusses; provide seat for window units.

Strip Window Units
Standard priority system for horizontal strip windows.

Figure 7.36   Building 8: details of the building shell.

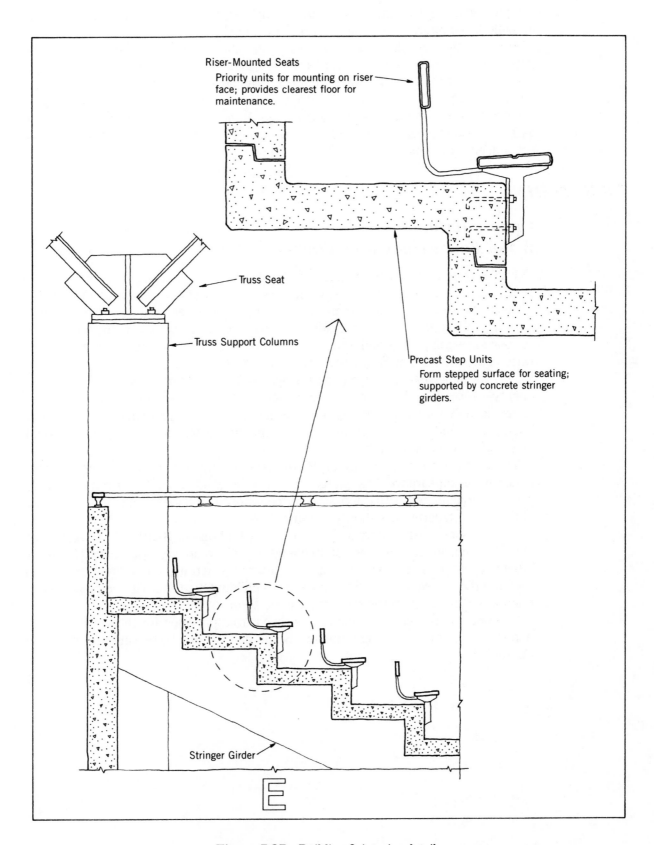

Riser-Mounted Seats

Priority units for mounting on riser face; provides clearest floor for maintenance.

Truss Seat

Truss Support Columns

Precast Step Units

Form stepped surface for seating; supported by concrete stringer girders.

Stringer Girder

E

Figure 7.37   Building 8: interior details.

complete and the planning could be totally independent. Some details for the development of the interior construction are shown in Figure 7.37.

Although the size, shape, general construction, and usage are all different, buildings 5, 8, and 9 are similar in that they are generally developed as single-space enclosures, with the interior construction largely free of the enclosure. The enclosure must generally accommodate the interior activities, but what happens on the inside is subject to considerable variation. Buildings 1 and 4 are quite different in this regard, while buildings 2, 3, 6, and 7 are transitional between the two situations—accommodating some limited interior variation.

7.9 BUILDING 9

Open Pavilion
Wood Pole Frame and Timber Truss

This is an open-air facility of medium span, consisting only of a large roof over a small amphitheater that is built into the ground (see Figure 7.38). Foundations and vertical supports for the roof structure are provided by wood poles with their bottom ends buried in the ground. The roof structure consists of a system of timber trusses supporting wood purlins and a wood plank roof deck. The roof surface is developed with a sheet metal roofing system with standing ribs.

This is a somewhat primitive structure, with some level of roughness in the finished quality—especially the timber poles, which may be simple tree trunks with only approximately round sections, tapered profiles with changing diameters, and many knots, splits, and so on. The truss members and purlins may also be rough, but could be standard structural lumber.

The roof deck is visible only on the underside, and thus only this surface is of concern for appearance. The plank units may be solid sawn, but are increasingly likely to be laminated, with the viewed surface chosen for appearance. General details of the wood structure are shown in Figure 7.39.

There is some analogy here with the structure for building 8, although the different materials, smaller size, and lack of an enclosing wall create considerable differences. What *is* common is the development of the interior with the exposed underside of the roof and the exposed structure. Also, as with building 8, the activity being housed is quite independent from the roof structure in terms of construction.

The only interior construction in this case is that provided for the seating, which defines the form of the amphitheater. Some details for the seating are shown in Figure 7.40.

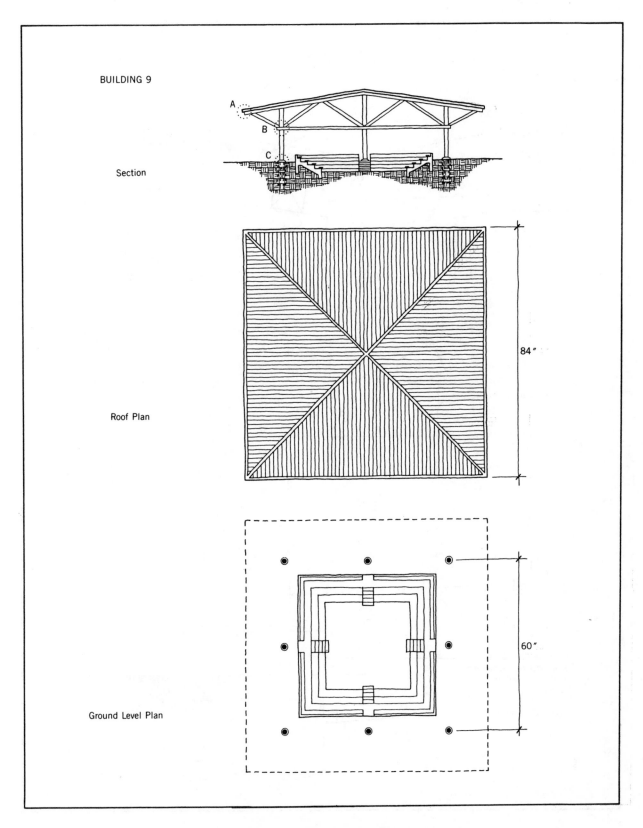

BUILDING 9

Section

Roof Plan

84″

Ground Level Plan

60″

Figure 7.38   Building 9.

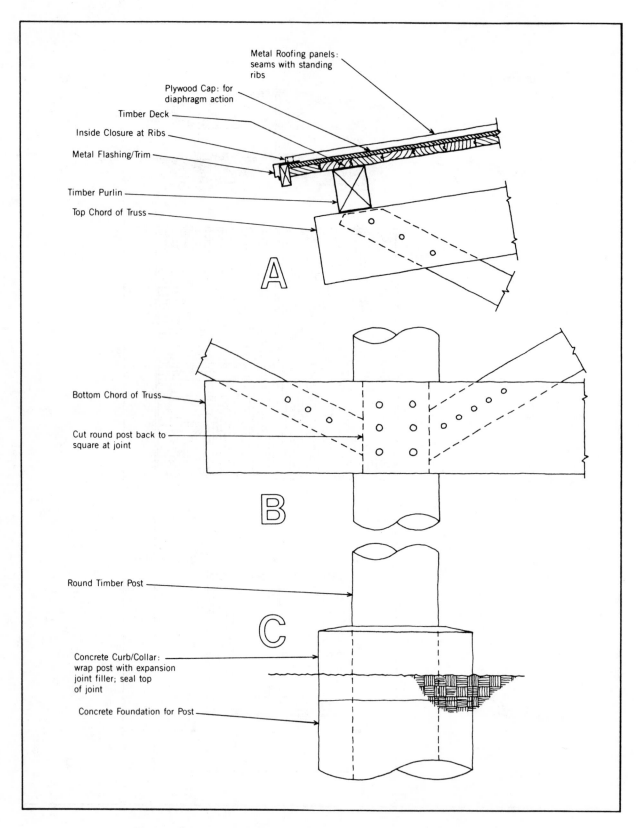

Figure 7.39 Building 9: details of the column and roof structures.

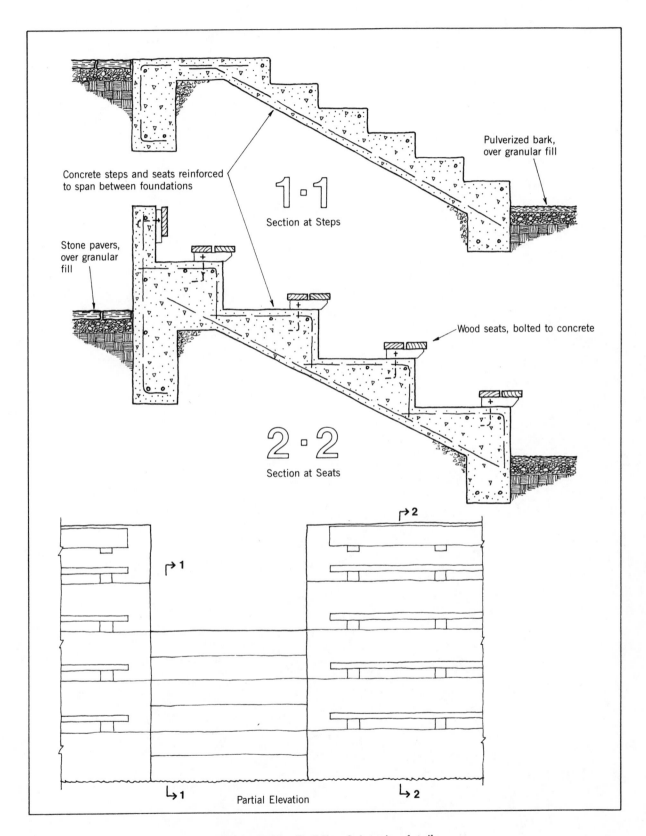

Concrete steps and seats reinforced to span between foundations

Pulverized bark, over granular fill

1·1

Section at Steps

Stone pavers, over granular fill

2·2

Section at Seats

Wood seats, bolted to concrete

Partial Elevation

Figure 7.40   Building 9: interior details.

Index